Cultural Histories, Memories and Extreme Weather

Extreme weather events, such as droughts, strong winds and storms, flash floods and extreme heat and cold, are among the most destructive yet fascinating aspects of climate variability. Historical records and memories charting the impacts and responses to such events are a crucial component of any research that seeks to understand the nature of events that might take place in the future. Yet all such events need to be situated for their implications to be understood.

This book is the first to explore the cultural contingency of extreme and unusual weather events and the ways in which they are recalled, recorded or forgotten. It illustrates how geographical context, particular physical conditions, an area's social and economic activities and embedded cultural knowledges and infrastructures all affect community experiences of and responses to unusual weather. Contributions refer to varied methods of remembering and recording weather and how these act to curate, recycle and transmit extreme events across generations and into the future. With international case studies, from both land and sea, the book explores how and why particular weather events become inscribed into the fabric of communities and contribute to community change in different historical and cultural contexts.

This is valuable reading for students and researchers interested in historical and cultural geography, environmental anthropology and environmental studies.

Georgina H. Endfield is Professor of Environmental History based at the University of Liverpool, UK. She is PI of an Arts and Humanities Research Council funded project exploring extreme weather events in UK history and is currently President of the International Commission for the History of Meteorology and Editor of *The Anthropocene Review*.

Lucy Veale is a Research Associate at the School of Geography, University of Liverpool, where she is working on the history of extreme weather events in the UK. Lucy's interests and expertise are in archival research in historical geography and she has worked on a wide range of projects relating to environmental, landscape and climate history.

Routledge Research in Historical Geography
Series Edited by Simon Naylor (School of Geographical
and Earth Sciences, University of Glasgow, UK) and
Laura Cameron (Department of Geography, Queen's
University, Canada)

This series offers a forum for original and innovative research, exploring
a wide range of topics encompassed by the sub-discipline of historical
geography and cognate fields in the humanities and social sciences. Titles
within the series adopt a global geographical scope and historical studies
of geographical issues that are grounded in detailed inquiries of primary
source materials. The series also supports historiographical and theoretical overviews and edited collections of essays on historical-geographical
themes. This series is aimed at upper-level undergraduates, research students and academics.

For a full list of titles in this series, please visit https://www.routledge.com/
Routledge-Research-in-Historical-Geography/book-series/RRHGS

Published

Historical Geographies of Prisons
Unlocking the Usable Carceral Past
Edited by Karen Morin and Dominique Moran

Historical Geographies of Anarchism
Early Critical Geographers and Present-Day Scientific Challenges
*Edited by Federico Ferretti, Gerónimo Barrera de la Torre, Anthony Ince
and Francisco Toro*

Cultural Histories, Memories and Extreme Weather
A Historical Geography Perspective
Edited by Georgina H. Endfield and Lucy Veale

Cultural Histories, Memories and Extreme Weather

A Historical Geography Perspective

**Edited by Georgina H. Endfield
and Lucy Veale**

Routledge
Taylor & Francis Group

LONDON AND NEW YORK

First published 2018
by Routledge
2 Park Square, Milton Park, Abingdon, Oxon OX14 4RN

and by Routledge
52 Vanderbilt Avenue, New York, NY 10017, USA

First issued in paperback 2020

Routledge is an imprint of the Taylor & Francis Group, an informa business

British Library Cataloguing-in-Publication Data
A catalogue record for this book is available from the British Library

Library of Congress Cataloging-in-Publication Data
A catalog record for this book has been requested

ISBN 13: 978-0-367-66768-9 (pbk)
ISBN 13: 978-1-138-20765-3 (hbk)

Typeset in Times New Roman
by codeMantra

Contents

Illustrations

Figures

Maps

Tables

Contributors

Georgina H. Endfield is Professor of Environmental History based at the University of Liverpool. Her research focuses specifically on climatic history and historical climatology, on human responses to unusual or extreme weather events, conceptualisations of climate variability in historical perspective and the links between climate and the healthiness of place. Much of her work has been concerned with colonial Mexico and nineteenth-century Africa, though for the past few years she has been working on various projects that focus on British climate history, and she is PI of an Arts and Humanities Research Council-funded project exploring historical extreme weather events in UK history.

James R. Fleming is the Charles A. Dana Professor of Science, Technology and Society at Colby College. He has earned degrees in astronomy (B.S. Pennsylvania State University), atmospheric science (M.S. Colorado State University) and history (Ph.D. Princeton University). He is a fellow of the American Association for the Advancement of Science and the American Meteorological Society and series editor of Palgrave Studies in the History of Science and Technology. His books include Meteorology in America, 1800–1870 (Johns Hopkins, 1990), Historical Perspectives on Climate Change (Oxford, 1998), The Callendar Effect (AMS, 2007), Fixing the Sky (Columbia, 2010) and Inventing Atmospheric Science (MIT, 2016).

Hywel M. Griffiths is a Senior Lecturer in Physical Geography in the Department of Geography and Earth Sciences at Aberystwyth University. His background is in fluvial geomorphology; in particular the rates, patterns and controls of river bed erosion. Alongside this interest he has developed research interests in a wide variety of fields, including integrating historical documentary records and geomorphological evidence to construct flood chronologies, remote sensing of river environments, river restoration and cultural geographies, but the relationship between people and rivers is always central to his work.

Alexander Hall is a historian whose work explores the role of science, the environment and belief systems in society. He is currently a research fellow at Newman University, Birmingham, where he is part of multi-disciplinary team exploring the social and cultural drivers perpetuating a clash narrative between science and religion. His previous research has explored the history of the British Met Office, flooding in the UK and memories of severe winters in the north-west of England.

Vladimir Janković is a Senior Lecturer at the Centre for the History of Science, Technology and Medicine, University of Manchester. His research focuses on scientific, cultural and social engagement with weather and climate since the 1700s. His work addresses the social origins of modern environmental medicine ca. 1750–1850 in connection to occupational and domestic exposures to bad air and bad weather; the historic and contemporary constructions of climate and weather, with the emphasis on the role of environmental crises in the development of 'disaster' research and policy process since WW2; twentieth-century work in small-scale meteorological research, urban climatology and micrometeorology; and the economics of weather and the private governance of climate change. He is author of *Reading the Skies: a cultural history of the English weather, 1650–1820* (Chicago 2000) in which he argues for the importance of a geographical turn in the history of meteorology and its role in observation, standardization and description of severe weather events. In *Confronting the Climate: British airs and the making of environmental medicine* (New York, 2010) he introduces domestic interiors as the prime site of the interaction between health and atmospheric environment. He also co-edited (with Christina H Barboza) *Weather, Local Knowledge and Everyday Life*. Rio de Janeiro, Brazil: MAST.

Cary J. Mock received his PhD at the University of Oregon in 1994 and has been a Professor at the University of South Carolina in the Department of Geography since 1999. His research interests are in historical climatology, Quaternary paleoclimatology, and synoptic climatology. His historical climate research includes studying weather extremes, such as hurricanes and cold air outbreaks, over the last several centuries. Recent research directions relate to how societies were impacted by weather and climate extremes. He has co-edited books on Historical Climate and Variability and Impacts in North America (2009), and Encyclopaedia of Quaternary Science (2013).

Ruth A. Morgan is an environmental historian and historian of science with a particular focus on Australia, the British Empire and the Indian Ocean. After completing her doctoral studies at The University of Western Australia, she joined the History program at Monash University in mid-2012. Her first monograph, *Running Out? Water in Western Australia*, was published in February 2015.

Cathryn Pearce, FRHists is a Visiting Lecturer and Research Fellow with the Department of History and the Greenwich Maritime Centre, University of Greenwich. Having previously published on the practice of wrecking—the plundering of shipwrecks by coastal inhabitants—she is now researching lifesaving and the role of coastal communities, focusing on the work of the Shipwrecked Fishermen and Mariners' Royal Benevolent Society. Pearce also serves on Council for the Society for Nautical Research, on the Editorial Board of the Mariner's Mirror (an INT1 journal) and with the British Commission for Maritime History.

Marie-Jeanne S. Royer has been, since December 2013, based in the Department of Geography and Earth Science at the University of Aberystwyth and working as Postdoctoral research associate on the AHRC funded project Spaces of experience and horizons of expectations: The Implication of Extreme Weather Events, Past, Present and Future. During this time she also taught on climate change and geohazards modules. Before coming to the UK, Marie-Jeanne completed a PhD on the "Interactions between TEK knowledge and climate change: the Cree of the Eastern James Bay, the Canada Goose and the Woodland Caribou". She also holds an MSc in geography on the "Socio-economic assimilation of the Mongols to the Han Majority in the Inner Mongolia Autonomous Region of the PRC". Since 2016 Marie-Jeanne is vice-chair of the International Geographical Union's (IGU) commission on Cold and High Altitude Regions (CHAR).

Eurig Salisbury is a Lecturer in Creative Writing in the Department of Welsh and Celtic Studies at Aberystwyth University and a researcher on the Cult of Saints in Wales project at the University of Wales Centre for Advanced Welsh and Celtic Studies. He has edited the work of a number of medieval Welsh poets (including Guto'r Glyn) and is also a published poet. Website: www.eurig.cymru.

Stephen Tooth is a Professor of Physical Geography in the Department of Geography and Earth Sciences at Aberystwyth University. His research interests focus on geomorphology and sedimentology, especially in the drylands of Australia and southern Africa. Particular research themes include: anabranching rivers; floodplains and floodouts; wetlands in drylands; channel-vegetation interactions; bedrock-influenced rivers; controls on gully erosion; long-term fluvial landscape development; palaeoenvironmental change and the use of drylands on Earth as analogues for Martian surface environments. He is also interested in environmental issues more generally, including current debates about global climate change and the Anthropocene and in science education.

Lucy Veale is Research Associate at the University of Liverpool. Lucy's expertise is in archival research in historical geography. She completed her PhD on the acclimatization of cinchona to British India in 2010 and has subsequently worked on a number of projects relating to environmental, landscape and climate history in the UK at the University of

Nottingham. She is currently working on 'extreme' weather as part of a multi-institution project funded by the UK's Arts and Humanities Research Council. This includes creating an extensive public database of historical extreme weather events in the UK, as well as exploring the impacts of the events, the human responses to them and the ways that they are remembered (or forgotten).

Ian Waites is a Senior Lecturer in the History of Art and Design at the University of Lincoln. He is particularly interested in the social and cultural history of post-World War Two Britain and is currently researching childhood memory and sense of place on a 1960s council estate in Lincolnshire. His most recent publication is an article on the disappearance of the estate's playgrounds, entitled "Once there were roundabouts", for *Uniformagazine* (Spring-Summer 2015). He turned 16 during the summer of 1976 and remembers doing very little during that time other than lying on his bed listening to Tubular Bells.

Acknowledgements

This volume is the result of a double session on Cultural Weather Memories held as part of the International Conference of Historical Geography, held at the Royal Geographical Society with the Institute of British Geographers, Kensington Gore, London, July 2015. The editors would like to thank all involved in this session and acknowledge the support of the Arts and Humanities Research Council [grant number AH/K005782/1].

1 Climate, culture and weather

Georgina H. Endfield and Lucy Veale

Introduction

Climate is at once local and global, cultural and scientific, political and popular. As an idea, it represents a "statistical construct, consisting of trends and averages that individuals can observe only indirectly" (Goebbert et al., 2012: 132; see also Spence et al., 2011), while discourses about climate have tended to be dominated by a contemporary global, scientific metanarrative of climate change. In recent years, however, it has been argued that climate can and should be localised, historicised and encultured (Hulme, 2012; Livingstone, 2012); should no longer just be the purview of the sciences (Hulme, 2009); and should be "understood, first and foremost, culturally" (Hulme, 2015: 1). In as much, there have been efforts to galvanise "new social understandings, criticisms and practices" of climate (Offen, 2014: 478).

Recent work, for example, has highlighted how the relationships between climate and culture appear everywhere in daily life, in clothing, the built environment, in social memories of past events, in emotions, adaptation, fiction, and in narratives of blame (Hulme, 2015: 1). There is now considerable interest in how different groups of people have conceptualised climate and have responded to its fluctuations (Strauss and Orlove, 2003; Boia, 2005; Behringer, 2010; Dove, 2014; Crate and Nuttall, 2016). A new way of thinking about climate, and so too "about and understanding the hybrid phenomenon of climate change" (Hulme, 2008: 6) has emerged, and there is a growing appreciation that what people make of climate and what they might do about it when it changes, "are complex cultural matters, with a matrix of narratives, specific meanings emerging in and from particular times and places" (Daniels and Endfield, 2009: 222).

To further facilitate a 're-culturing' of climate, it has been suggested that we might "think more directly about weather" (Hulme, 2015: 2). As de Vet (2013: 198) has argued, "in terms of everyday human experience, climate and long term climate change takes expression through specific local weather patterns". Weather provides the lens through which the relationship between culture and climate is most easily viewed (Hulme, 2008). It can of course also be experienced, monitored, observed, culturally mediated, and has an "immediacy and evanescence" that climate does not (Hulme, 2015: 3). Weather is also very local (Tredinnick, 2013). It sets the background to our lives; it can provide a context, can affect character, and can be used symbolically in stories (Hulme, 2015: 1–2). Yet, there is still a relative lack of research that explores "the interconnections [and] intricacies" of weather's imprint on societal lifeways (de Vet, 2013: 198). Moreover, with some exceptions (Pillatt, 2012; Hall and Endfield, 2015), only limited attention has been paid to the 'interconnections' and 'imprints' associated specifically with extreme or unusual weather, such as droughts, floods, storm events and unusually high or low temperatures.

There is particular concern over the impacts of such events, for while social and economic systems have generally evolved to accommodate some deviations from 'normal' weather conditions, this is rarely true of extremes. For this reason, these events can have significant and immediate repercussions. Scholarship has tended to focus on the social and economic implications of extreme events (for example Diaz and Murnane, 2008), and more recently, there have been a number of interventions considering the connections between extreme weather and climate change (Hulme, 2014). In this volume, however, we are concerned more with the cultural contingency of such events, on the importance of cultural context in understanding their effects, on the interpretations, articulations and inscription of unusual and extreme weather. We are also interested in the mechanisms through which these events are recorded, recalled, obscured or subsumed by other events. Such histories and memories play a significant role in the popular understanding and articulation of current debates about weather and climate. We bring together scholars whose research on these themes draws on a range of original unpublished archival materials, historical meteorological accounts and registers, newspaper archives, and oral history approaches to investigate cultural histories of extreme weather in a range of contexts and spaces. Through their consideration of distinctive case studies from the UK, Canada, the US, continental Europe and Australia, and based on both land and sea, the authors explore the ways in which different types of unusual and extreme weather have been experienced, perceived and recorded in different contexts throughout history, highlighting how some events become culturally inscribed at the 'expense' of others.

The chapters address a number of key framing questions: how and why are particular weather events remembered while others are forgotten? How are weather events recorded and recounted? How have particular weather events become inscribed into the cultural fabric or embedded in environmental knowledge over time? Our authors refer to varied forms of recording and remembering weather and look at how these together act to curate, recycle and transmit knowledge of extreme events across generations. They speak to the different scales and forms of memory-collective, popular, public and counter memory (Glassberg, 1996). They also demonstrate how geographical context, particular physical conditions, an area's social and economic activities and embedded cultural knowledge, norms, values, practices and infrastructures can affect community experiences of and responses to unusual weather. Adopting Hulme's terms, all chapters represent a "downscaling" of weather, taking account of location and contingency of place in understanding the physical and mental imprint of weather events (Hulme, 2008: 8).

In order to introduce and situate the chapters, what follows is a brief overview of approaches to and recent work in climate and weather history scholarship, organised by themes that have helped shape the current research and to which these chapters make a significant contribution.

Finding weather: elemental life, ubiquity and the recording of extremes

The weather is ubiquitous. It has been woven into human experience and the cultural memory and fabric of communities through oral histories, proverbs, folklore, narrative and everyday conversation (Strauss and Orlove, 2003). Weather shapes, changes and defines us, and "we are who we are, indirectly and directly, because of the weather we lead our lives in" (Tredinnick, 2013: 15). For this reason, 'registers' of weather and climate, as Hulme (2008: 7) suggests, can be read "in memory, behaviour, text, and identity as much as they can be measured through meteorology". Such non-numerical forms of testimony represent vital media through which information about short-lived weather events, as well as long-term climate change, is gathered and transmitted across generations.

In an 1958 article published in *Weather* (the journal he helped to establish in 1946 while he was President of the Royal Meteorological Society), British climatologist, Gordon Manley, argued that "in a modern civilisation, the existence of a public memory of the weather is essential" (Manley, 1958: 11). This memory takes many forms. Climate and weather have long been the subjects of private narratives, diaries, chronicles and sermons dating back to the later seventeenth and eighteenth centuries (Janković, 2000), and a very diverse group of people were involved in observing and recording weather in this period, either in networks or independently, including gentlemen savants, physicians, sea captains, clergy naturalists, university professors and travellers (Daston, 2008). In the earliest records, emphasis was often placed on "qualitative and narrative framing of 'important' weather", or 'meteoric' weather – the unusual or extreme event that disrupted everyday life at the local level (Janković, 2000: 9). The "notoriously local and changeable" nature of the weather, however, served to stimulate a motivation to identify some kind of order among the range of "fickle meteorological phenomena" (Daston, 2008: 234–235), and from the mid to late eighteenth century, more quotidian recording practices were adopted, whereby people collated daily weather reports of their local weather, supplemented by readings from meteorological instruments (Golinski, 2003).

Extremes continued to be viewed as noteworthy, however, and many sources contain detailed descriptions of such events and their implications, including from the 1700s local, regional and national newspapers (Grattan and Brayshay, 1995; Gallego et al., 2008). While all our authors draw on a wide range of source materials, including meteorological registers, personal diaries and various forms of narrative account, many of the chapters in this volume make use of newspaper reports, institutional records and bulletins. Ian Waites' study of the 1976 heatwave and drought in the UK, for example, draws on content from both national and local newspapers, while Cathryn Pearce's study of shipwrecked fishermen and mariners, and the emergence of benevolent societies in Southwest England, uses materials from local chronicles and weekly gazettes. US-based newspaper reports are central to Cary Mock's research into historical 'hurricane memory', and Alexander Hall makes use of local and regional newspapers in his work exploring the aftermath and commemoration of the 1953 East Coast floods. Newspaper accounts are also central to Ruth Morgan's study of the cultural memory of the 1914 drought in Western Australia.

Alongside these direct qualitative and quantitative sources of weather information, there are many other sources of information on historical weather events, including travel accounts, legal documents, crop and tax records, maps, paintings, etchings, plans and images, and collections of correspondence. Many folk traditions express "knowledge regarding the full range of weather and climatic conditions" (Strauss and Orlove, 2003: 7), based on familiarity with long-term climate variability. Extreme weather has also become embedded in the design and construction of vernacular buildings, drainage systems, clothing and alterations to modes of transport (Hassan, 2000). Extreme events that resulted in trauma, such as floods, can result in epigraphic recording of such events, as well as other forms of commemoration. Figure 1.1, for example, shows a set of epigraphic flood markers carved into the wall at the side of the River Trent, near Trent Bridge, in Nottingham, England, demonstrating the way in which particular flood events are remembered and compared. Alexander Hall highlights how similar flood markers at the entrance to St Margaret's Church in Kings Lynn, the inspiration behind his study, are among the ways in which local weather events can in effect be preserved, as a form of community memory.

The weather is frequently represented or invoked in literature and has been associated with human action and mood in novels, poems and plays. Good, bad or 'unsettled' weather provides a backdrop or context for a storyline or is used metaphorically to "intensify the reader's understanding of emotional states and social conditions" (Collins, 2013: 2). Yet the experience of weather, particularly unusual weather, can also be the subject of purposeful representation. It follows that the impact of weather events on human experience can be traced through literary media. Griffiths and colleagues focus on a medieval poem by Lewys Glyn Cothi to investigate how this form of narrative can provide insight not only into particular flood events in the

Figure 1.1 Flood markers, Trent Bridge, Nottingham.
Photo: Mark Bentley.

early modern period but also affords a balanced view of flooding as a part of everyday life in what they term a 'hydrographic culture', a society that kept records of their relationship to local hydrology. The poem represents a form of 'flood heritage' and is a valuable source for anyone interested in the environmental impacts of extreme flood events, including on trees, and in the historic practice of assigning human characteristics to rivers and their behaviour.

Tapping into local weather memories through interview and oral history work can yield very useful information on perceived changes in such unusual events, their frequency and intensity, and the impacts and responses they engendered, as well as revealing how people conceptualise and contextualise the risks of any future events (Leyshon and Geoghegan, 2012). Short, retrospective, interview approaches have proved to be useful for establishing how people are able to generate valuable autobiographical experiences of weather and the everyday (de Vet, 2013) as well for as retrieving memories. Moreover, such personalised weather narratives have a significant role to play in popular understanding and articulation of debates about weather and climate (Lejano et al., 2013). Marie-Jeanne Royer adopts this approach in her work to explore intergenerational ecological knowledge and perceptions of a changing climate based on weather memory and wisdom among Cree communities in Canada, while Ruth Morgan highlights the utility of pre-existing oral histories for revealing the deeply personal nature of the drought response in early-twentieth-century Western Australia and the contested cultural memory therein.

Social networking and sharing platforms like Twitter, Facebook, Flickr and YouTube offer opportunities for innovative public geographies (Kitchin, 2013), including historical ones. These kinds of digital approaches not only "open up... new forms of representation... and expanded publics" (Yusoff and Gabrys, 2011: 519) but also represent the repositories of future weather memories. Contemporary instances of unusual or extreme weather are now perhaps more likely to be captured and recorded through webpage entries, blog narratives and tweets as they are in newspaper reports or other written records. These forms of social media, however, can also provide a medium through which to assemble memories of past weather events. Ian Waites' chapter on the 1976 heatwave, for example, makes extensive use of online reminiscences of the event, demonstrating the rich empirical information on weather memory that is retrievable from blogs and other web entries, and highlights the value of online repositories as examples of future cultural memory.

The waxing and waning of weather memory: remembering (and forgetting) weather events

The chapters in this volume all point to different forms of weather memory but also consider how those weather memories have been made. Hywel Griffiths and colleagues, for example, discuss the 'encoding' of weather memory, which is subsequently transmitted across generations through "a complex series of occurrences" – written and performed, collected and curated. Memory tends to be distorted with respect to more recent extreme weather events, according to what Harley (2003) refers to as the "recency effect". Indeed, as Eden (2008: 4) has suggested, with the exception of the most extreme or unusual events, "once a weather phenomenon has reached two years old it seems to fall out of the human memory bank". In a UK context, for example, exceptionally severe winters, such as 1947 and 1962–1963, and indeed the summer of 1976, though recognised to be extreme, appear also to have claimed priority in people's memories as idealised stereotypes of seasonal conditions. Such events are often regularly considered to be 'unprecedented', the memory of similar events in the past having effectively been overwritten by a more recent event.

This issue of 'forgetting' is raised in Cary Mock's chapter on US hurricane memory. Memory of these events, he argues, falls into three categories: persistent memory, intermittent memory and lack of memory. Storms associated with persistent memory, and which are defined as those having a significant legacy, are memorialised largely because they affected populous areas, causing loss of life and widespread damage. Those events associated with intermittent memory tend to be those that are forgotten but rediscovered numerous times during anniversaries or in association with education

and outreach efforts. Storms associated with a lack of memory, by definition, are challenging to interpret, but likely relate to those events that affected rural or remote areas or perhaps caused little or negligible damage.

Particularly dramatic or extreme events or those that resulted in major
disruptions tend to seize popular attention and "evoke strong feelings, making them memorable and, therefore, often dominant in processing" (Marx
et al., 2007: 48). There is also a tendency to remember events from the distant past with greater clarity if related to unusual weather. In this respect,
weather provides a metacognitive role in organising memory (Harley, 2003),
and unusual or extreme weather events can act as "anchors for personal
memory" (Hulme, 2009: 12). Indeed, while we *make* history in normal
weather (Tredinnick, 2013), extreme weather *shapes* history. Memories of
past weather events, however, are not always bad memories. As Ian Waites
demonstrates in his chapter, people who lived through the summer of 1976
have tended to express enjoyment in remembering it, especially those who
were children at the time or for whom the prolonged heat coincided with key
life events such as marriage, childbirth, new jobs or the purchase of a new
vehicle. Theirs was a shared, positive nostalgia.

Alex Hall's chapter highlights the important role that key institutions,
such as the church, have played in the immediate aftermath of extreme
weather events and in the intergenerational commemoration of these events.
Echoing work in public history (see Anderson, 1991; Glassberg, 1996), this
work illustrates how the 1953 flood event has in effect contributed to a shared
sense of cultural memory and has been remembered 'in common' but a very
similar meteorological event in 1978 less so. In Cathryn Pearce's chapter, institutional memory of weather events is created through the record-keeping
of the Shipwrecked Fishermen and Mariner's Benevolent Society, a charitable organisation established following the storm of 28 September 1838 in
which 26 men lost their lives, while Marie-Jeanne Royer's chapter highlights
how weather memory is 'stored', curated and used as part of an unwritten
traditional ecological knowledge in Cree culture.

Relational context informs people's understandings and expectations of
weather (Hulme, 2009; Leyshon and Geoghegan, 2012) allowing for a "min-
gling of place, personal history, daily life, culture and values" (Brace and
Geoghegan, 2011: 289). Whether direct or received, therefore, it follows that
weather memory and weather knowledge are usually necessarily situated.
Although memory is spatially constituted and attached to key 'sites', both
physical such as local landmarks, and non-material 'sites' such as celebrations, traditions or folklore (Nora, 1989; Hoelscher and Alderman, 2004),
weather memory *per se* is also a function of and shapes place. Place in
turn has an important bearing on the weather and how weather variability and weather events are experienced and recalled. Moreover, as Pillatt
(2012: 34) has argued, "the weather in which one stands can be as much
responsible for generating sense and use of place as the ground in which one
stands". Place thus plays a central role in influencing weather histories and

memories, and weather provides a frame of reference that also helps to situate other memories, while in turn contributing to the making and meaning
of place. Ian Waites highlights how particular spaces – notably the domestic
garden – emerge as prominent for remembering the particular adaptations
that were made to cope with the 1976 heatwave, as more time was spent
living and indeed sleeping outside. In contrast, in Alex Hall's chapter, the
church emerges as a key site of response to the 1953 East Coast flood event,
as an institution central to community, to the maintenance of social cohesion and to everyday routine through worship and business, while it played
a diminished role by the time the 1978 flood event took place. The church
also served to normalise the flood experience, by temporally placing and
situating the 1953 flood event within a longer history of repeat floods.

Small talk and turning points: framing and apprehending weather events

Climate has become 'big talk'. It has become politicised, part of an econo-
metric discourse, while weather remains "small talk" (Tredinnick, 2013: 13)
in that it is a relatively easy, often apolitical, conversational topic, "avoiding
personal or sensitive matters" (Harley, 2003: 103; Strauss and Orlove, 2003).
By extension, talking about weather *memories* seems to be similarly 'easy'.
Ian Waites' chapter, for instance, is a perfect illustration of the characteris-
tically British preoccupation of talking (and usually complaining) about the
weather and demonstrates the way in which particular events can come to
dominate weather memory 'talk'. Looking back on 'golden summers' in the
UK and specifically the record temperatures of the summer of 1976, Waites
recalls how people talk about what became a national and prominent event
in the country's weather history, noting the very local stories, experiences
and meteorological records, including dry-weather-themed fancy dress!

  People tend to draw on past personal experience to make sense of the current and possible effects of the weather. Current situations are often related
to memories of previous apparently analogous events or others' experiences,
what we might refer to as 'received memories' (Marx et al., 2007: 48). People do not have to have a lived experience of past weather, however, to feel
nostalgia for it (Chase and Shaw, 1989). Baker and Kennedy (1994) draw
a distinction between 'real' nostalgia for some remembered previous time
and a so-called 'stimulated' nostalgia – a form of vicarious nostalgia evoked
from secondary sources, images or received wisdoms (Goulding, 1999: 179).
Moreover, as Forgas et al. (2009: 254) argue, "remembering the details of
everyday scenes is a fragile process that is often influenced by what people
pay attention to, as well as contamination by subsequent, incorrect information" (see also Fiedler et al., 1991; Wells and Loftus, 2003). Misleading information after an event can produce a false memory later on – the so-called
'misinformation effect' (Schooler and Loftus, 1993). Human perceptions
"are prone to selective retention and biased assimilation of information"

(Goebbert et al., 2012: 133), whether this is accidental or deliberate. Ruth Morgan, for example, draws on a set of oral histories conducted in the later part of the twentieth century, many of which reveal deeply personal, individual reflections on the 1914 drought in Western Australia. These are infused with what Morgan calls "pioneer mythology", part of a narrative of resilience and adaptation that in Stannage's (1985) terms represents a "gross distortion of the reality of the past" (p. 44 this volume) not least because, as discussed below, it overlooks other more poignant stories of loss and deprivation.

As Cronon (1992: 1347) has highlighted, the past is filled with "many stories, of many places in many voices, pointing towards many ends". Historical events are assembled into causal sequences and narratives that "order and simplify those events to give new meanings" to the past such that inevitably "we divide the causal relationships of an ecosystem with a rhetorical razor that defines included and excluded, relevant and irrelevant, empowered and disempowered" (Cronon, 1992: 1349). Particular aspects of a historical event and by extension a historical weather event can thus be emphasised, elaborated, obscured, omitted or ignored in order to tell a particular story. The impact of extreme weather varies between individuals, depending on a multitude of factors, which are in turn informed by environmental context and cultural and historical experience (MacDonald et al., 2010; Endfield, 2014). Yet weather events can be understood and 'framed' by and apprehended through very different narratives. There are often multiple, diverse, contested and competing narratives, narratives of ascension and declension, or hope and despair, privileging predominantly cultural or environmental explanations of change and transition depending on the particular frames or tropes that are used to tell a story.

In his study of the Norfolk heatwave of the turn of the twentieth century, Hulme (2012: 8) explored "the different ways in which a specific climate event 'lives on' and becomes a resource that is used in different cultural pursuits, social realities and scientific enterprises", with a view to offering a contingent view of the event. As this volume reveals, other weather events can be considered similarly, with key events being understood and articulated in very different ways, with different frames of reference. Weather events give rise to different, sometimes contested, cultural imaginaries at different points in time, though all serve to memorialise the event. Ruth Morgan's chapter on drought in Western Australia, for example, highlights several stages of climate apprehension and different frames of reference. There was initially an appropriation of the apparently benign average climate conditions of the region in the promotional strategy adopted by Western Australia's first Premier to encourage settlement of the region. The drought in 1914 challenged this but was framed equally as "aberration and opportunity" – at once a 'set back' in Western Australia's progress and prosperity and a key life event around which pioneer identities were created and shaped. Morgan also reveals the importance of contested cultural weather memories and highlights how, for Aboriginal people of Southwest Australia, the drought was a major turning point with respect to territorial dispossession.

Contested weather memory is also considered in Hall's chapter. He notes, for example, how national media and governmental narratives framed the tragic events of late January/early February 1953 as 'exceptional', setting in motion a 'wartime spirit'. Community-level narratives, while echoing this sense of community resilience and cohesion, however, speak to a normalisation of the event within a much longer history of flooding, rather like the hydrographic culture to which Griffiths et al. refer in their chapter.

Even if events are frequent occurrences, and hence normalised in a particular culture or place, some events do seem to act as temporal 'benchmarks' against which others are compared. Morgan, for example, considers the lasting memory of the 1914 drought in Western Australia as an event against which later dry periods were compared, thereby providing insight into its role in teaching and learning and shaping responses to contemporary and future droughts. Framed as a severe event but also something of an anomaly, far from the norm, the event was appropriated as a historical benchmark. The same is true of a number of the hurricanes considered in Cary Mock's chapter. As he points out, however, in this particular context and considering this particular type of weather, it is those events that led to most damage that are most commonly considered benchmark events. Marie Jeanne Royer's chapter demonstrates how changes in the populations of certain key bird and animal species, which provide a livelihood for Cree hunters, may also provide temporal markers of changing weather and environmental conditions, though it is very difficult to disentangle the cause or causes of such changes. Major meteorological events, particularly those that leave discontinuities in the historical record, do not necessarily require major causes, so much as a suite of social, economic, political, demographic and environmental factors that have the potential to coalesce at a particular point in time to cause dislocation or stimulate some form of societal or environmental change (Coombes and Barber, 2005).

Drawing links between weather events and social or general environmental change in history, however distant or recent, is thus somewhat challenging. Nevertheless, several chapters in this volume highlight how particular events, and memories of these events, are invoked to explain a period of transition, a change in policy or innovation. The severe gale of 28 September 1838, a storm that had a devastating impact on the fishing communities of North Devon in the Southwest of England and the story that opens Cathryn Pearce's chapter, was in fact the direct stimulus for the establishment of the Shipwrecked Fishermen and Mariners Society (SMS), a subscription-funded society focused on relieving the extreme weather-related suffering of fishermen, mariners and their dependent families. Similarly, Alex Hall explains how the 1953 flood led to the development and implementation of a more efficient storm warning system, sea defences and disaster planning. These initiatives, which could be considered adaptive strategies, were in turn hailed as the reason that later events, including that of 1978, had less significant impacts.

Other events, however, appear to have been retrospectively reframed as key turning points in social or indeed climate history. Ian Waites, for example, shows how the heatwave of 1976 triggered a sense of happiness and community spirit, forging an imagined community through which disparate groups and individuals were brought together as a collective as a result of a shared, net positive, experience (Glassberg, 1996). The effects of the heatwave not only contributed to a markedly slower way of life but also coincided with a time when social inequalities were lessened and new social horizons opened in a form of 'cultural enlightenment'. Yet as Waites explains, contemporary reflections suggest that this was also a time of social transition. As one of his sources notes, for example, "this was a time that got people sitting outside pubs, having BBQs in the back gardens, things that are the norm now" (p. 25 this volume). The 1976 heatwave, however, also came after the driest 16-month period in 250 years, such that the summer months aggravated rather than caused major water shortages across the country. It raised awareness of the ageing water infrastructure, highlighted the implications of water scarcity for wellbeing and began to alert people to the more sinister 'global problem' of climate change.

Making use of cultural weather memory

As Mike Hulme (2015: 9) has asserted, climate has a cultural history. As "the primary sign of the inextricability of culture and nature" (Bate, 1996: 437), however, weather too has a cultural history. The chapters in this volume reveal how this particular history has been articulated through written documents, both printed and hand written, through inscriptions, poems and oral histories and traditional ecological knowledge systems. All represent in one way or another evidence of cultural weather memory.

Collectively, the chapters speak to the different mechanisms and media through which weather events have been understood and interpreted and how these events, and experiences of them, have shaped and framed people's lives (Hulme, 2015; Tredinnick, 2013). In de Vet's (2014) terms, the imprints of normal weather in everyday life might be termed 'weather ways', representative of a day-to-day existence with weather, an essentially elemental way of life. The authors reveal that extreme or unusual weather events act to disrupt these 'weather ways', both positively and negatively. While such events often lead to negative experiences of loss, upheaval and deprivation, they can also draw people together, fostering a sense of community and collective cohesion, whether this is through key institutions, such as the church and charitable organisations, a sense of belonging through traditional ecological knowledge or simply through the collective, shared experience of living through an event. There are also contested weather memories and different ways events have been framed and apprehended as good or bad, contemporaneously or retrospectively. Cultural weather memory is thus at once celebratory, commemorative, contemplative and contested.

The chapters in this volume also present compelling stories of weather events and memorialisation, sometimes melancholy, declensionist and negative, sometimes more nostalgic, upbeat and optimistic, and they provide fascinating insights into cultural, social and environmental change. Such stories may also offer "political possibilities" (Cameron, 2012), serving a real and very tangible purpose. The construction of regionally specific climatic and weather histories (including extreme event histories) forms a crucial component of any research that seeks to understand the nature of events that might take place in the future. Thinking with the kinds of stories and memories featured in this volume may provide insight into how different communities in different contexts might be affected by, comprehend and respond to future events and may thus help us in terms of translating cultural weather memories and stories into climate futures. Recent scholarship, for instance, is beginning to acknowledge that not only advances in science but also local observations and experiences can prove useful in helping to conceptualize and understand how weather and its variations affect and have affected people at the local level (Lazrus and Peppler, 2013). As Vannini et al. (2012: 363) have argued, "the ways in which people experience and talk about weather... and the ways they sense and comprehend meteorological processes and draw significance from them are not only interesting but also particularly valuable keys to deciphering larger scale processes". Histories, memories, and vivid, vicarious or actual experiences of weather events, may thus serve a powerful role in the judgement and popular understanding of both local and global climatic change (Nisbett and Ross, 1980, cited in Marx et al., 2007: 49).

There are, of course, myriad interpretations of what has happened to, and is happening to, our climate, which are based on personal or familial experience, memories, nostalgia, information derived from the media, meteorological organisations and scientific documentation. Indeed, Goebbert et al. (2012) conclude that perceptions of local weather events and climate change more broadly inevitably follow the heuristic nature of public knowledge, the implication being that personal knowledge or experience plays a less informative role in shaping people's perspectives on climate changes than say external, prevailing scientific knowledge. Yet engagement with and communication of climate change risk is thought to be more effective and appropriately targeted if it takes into account relevant personal and vicarious experiences in the form of stories, narrative, memories and anecdotes of the kind considered in this volume (Marx et al., 2007). Such stories of weather extremes and the way that they become inscribed into cultural weather memory, therefore, may have the "power to take us beyond the usual academic tradition of narrative analysis and to put us in contact with valuable resources for moral thought and action" (Morris, 2001: 56).

References

Akerlof K, Maibach EW, Fitzgerald D, Cedeno AY and Neuman A (2013) Do people "personally experience" global warming, and if so how, and does it matter? *Global Environmental Change*, 23 (1): 81–91.

Anderson B (1991) *Imagined Communities. Reflections on the Origins and Spread of Nationalism*. New York: Verso.

Baker SM and Kennedy PF (1994) Death by nostalgia: a diagnosis of context-specific cases. *Advances in Consumer Research*, 21 (1): 169–176.

Bate J (1996) Living with the weather. *Studies in Romanticism*, 35 (3): 431–447.

Behringer W (2010) *A Cultural History of Climate*. Cambridge: Polity Press.

Boia L (2005) *The Weather in the Imagination*. London: Reaktion Books.

Brace C and Geoghegan H (2011) Human geographies of climate change: landscape, temporality and lay knowledge. *Progress in Human Geography*, 35: 284–302.

Cameron E (2012) New geographies of story and storytelling. *Progress in Human Geography*, 36: 573–592.

Chase M and Shaw C (1989) The dimensions of nostalgia. In Shaw C and Chase M (eds.) *The Imagined Past: History and Nostalgia*. Manchester: Manchester University Press: 1–17.

Collins L (2013) The frosty winters of Ireland: poems of climate crisis, 1739–41. *The Journal of Ecocriticism*, 5 (2): 1011.

Crate SA and Nuttall M (eds.) (2016) *Anthropology and Climate Change: From Encounters to Actions*. Oxford: Routledge.

Cronon WA (1992) A place for stories: nature, history, and narrative. *The Journal of American History*, 78 (4): 1347–1376.

Coombes P and Barber K (2005) Environmental determinism in Holocene research: causality or coincidence? *Area*, 37: 303–311.

Daniels S and Endfield GH (2009) Narratives of climate change: introduction. *Journal of Historical Geography*, 35 (2): 215–222.

Daston L (2008) Unruly weather: natural law confronts natural variability. In Daston L and Stolleis M (eds.) *Natural Law and Laws of Nature in Early Modern Europe: Jurisprudence, Theology, Moral, and Natural Philosophy*. Aldershot: Ashgate: 233–248.

de Vet E (2013) Exploring weather-related experiences and practices: examining methodological approaches. *Area*, 45: 198–206.

de Vet E (2014) *Weather-ways: experiencing and responding to everyday weather*. Unpublished PhD thesis. Wollongong, NSW: University of Wollongong.

Diaz HF and Murnane RJ (2008) *Climate Extremes and Society*. Cambridge: Cambridge University Press.

Dove MR (2014) *The Anthropology of Climate Change: An Historical Reader*. Chichester: Wiley-Blackwell.

Eden P (2008) *Great British Weather Disasters*. London: Continuum Books.

Endfield GH (2014) Exploring particularity: vulnerability, resilience, and memory in climate change discourses. *Environmental History*, 19 (2): 303–310.

Fiedler K, Asbeck J and Nicke S (1991) Mood and constructive memory effects on social judgement. *Cognition and Emotion*, 5 (5–6): 363–378.

Forgas JP, Goldenberg L and Unkeach C (2009) Can bad weather improve your memory? An unobtrusive field study of natural mood effects on real-life memory. *Journal of Experimental Social Psychology*, 45 (1): 254–257.

Gallego D, García-Herrera R, Prieto R and Peña-Ortiz C (2008) On the quality of climate proxies derived from newspaper reports. A case study. *Climate of the Past*, 4 (1): 11–8.

Glassberg D (1996) History and the study of memory. *The Public Historian*, 18 (2): 7–23.

Goebbert K, Jenkins-Smith HC, Klockow K, Nowlin MC and Silva CL (2012) Weather, climate, and worldviews: the sources and consequences of public perceptions of changes in local weather patterns. *Weather, Climate, Society*, 4: 132–144.

Golinski J (2003) Time, talk, and the weather in eighteenth-century Britain. In Strauss S and Orlove B (eds.) *Weather, Climate, Culture*. Oxford: Berg: 17–38.

Goulding, C (1999) Heritage, nostalgia, and the "grey" consumer. *Journal of Marketing Practice: Applied Marketing Science*, 5 (6/7/8): 177–199.

Grattan J and Brayshay M (1995) An amazing and portentous summer: environmental and social responses in Britain to the 1783 eruption of an Iceland volcano. *The Geographical Journal*, 1: 125–134.

Hall A and Endfield GH (2016) Snow Scenes: exploring the role of memory and place in commemorating extreme winters. *Weather, Climate and Society*, 8: 5–19.

Harley TA (2003) Nice weather for the time of year. The British obsession with the weather. In Strauss S and Orlove BS (eds) *Weather, Climate, Culture*. London: Berg: 103–118.

Hassan F (2000) Environmental perception and human responses in history and prehistory. In McIntosh RJ, Tainter JA and McIntosh SK (eds.) *The Way the Wind Blows: Climate, History and Human Action*. New York: Columbia University Press: 121–140.

Hoelscher S and Alderman DH (2004) Memory and place: geographies of a critical relationship. *Social and Cultural Geography*, 5: 347–355.

Hulme M (2008) Geographical work at the boundaries of climate change. *Transactions of the Institute of British Geographers*, 33: 5–11.

Hulme M (2009) *Why We Disagree About Climate Change*. Cambridge: Cambridge University Press.

Hulme M (2012) Telling a different tale: literary, historical and meteorological readings of a Norfolk heatwave. *Climatic Change*, 113 (1): 5–21.

Hulme M (2014) Attributing weather extremes to 'climate change' a review. *Progress in Physical Geography*, 38 (4): 499–511.

Hulme M (2015) Climate and its changes: a cultural appraisal. *Geo: Geography and Environment*, 2 (1): 1–11.

Janković V (2000) *Reading the Skies: a Cultural History of English Weather, 1650–1820*. Manchester: Manchester University Press.

Kitchin R (2013) Big data and human geography opportunities, challenges and risks. *Dialogues in Human Geography*, 3 (3): 262–267.

Lazrus H and Peppler R (2013) Ways of knowing: traditional knowledge as key insight for addressing environmental change, theme introduction. *Weather, Climate and Society* special collection (2013) http://journals.ametsoc.org/page/Ways (accessed 20 November 2013).

Lejano R, Tavares-Reager J and Berkes B (2013) Climate and narrative: environmental knowledge in everyday life. *Environmental Science & Policy*, 31: 61–70.

Leyshon C and Geoghegan H (2012) Landscape and climate change. In Thompson HPI and Waterton E (eds.) *The Routledge Companion to Landscape Studies*. London: Routledge.

Livingstone DN (2012) Reflections on the cultural spaces of climate. *Climatic Change*, 113: 91–93.

Macdonald N, Jones CA, Davies SJ and Charnell-White C (2010) Historical weather accounts from Wales: an assessment of their potential for reconstructing climate. *Weather*, 65 (3): 72–81.

Manley G (1958) The great winter of 1740. *Weather*, 3 (1): 11–17.

Marx SM, Weber EU, Orlove BS, Leiserowitz A, Krantz DH, Roncoli C and Phillips J (2007) Communication and mental processes: experiential and analytic processing of uncertain climate information. *Global Environmental Change*, 17: 47–58.

Morris D (2001) Narratives, ethics and pain: thinking with stories. *Narrative*, 9 (1): 55–77.

Nisbett RE and Ross L (1980) *Human Inference: Strategies and Shortcomings of Social Judgment*. Englewood Cliffs, NJ: Prentice-Hall.

Nora P (1989) Between memory and history: les lieux de memoire. *Representations*, 26: 7–24.

Offen K (2014) Historical geography III. *Progress in Human Geography*, 38 (3): 476–489.

Pillatt T (2012) Experiencing climate: finding weather in eighteenth century Cumbria. *Journal of Archaeological Method and Theory*, 19: 564–581.

Schooler JW and Loftus EF (1993) Multiple mechanisms mediate individual differences in eyewitness accuracy and suggestability. In Puckett JM and Reese HV (eds.) *Mechanisms of Everyday Cognition*, NJ: Lawrence Erlbaum Associates: 177–204.

Spence A, Poortinga W, Butler C and Pidgeon NF (2011) Perceptions of climate change and willingness to save energy related to flood experience. *Nature Climate Change*, 1 (1): 46–49.

Stannage T (1985) Western Australia's heritage: the pioneer myth. *Studies in Western Australian History*, 29: 145–56.

Strauss S and Orlove BS (2003) *Weather, Climate, Culture*. Oxford: Berg.

Tredinnick M (2013) The weather of who we are. An intimate essay on the weather, the self and Australianness. *World Literature Today*, 87 (1): 12–15.

Vannini P, Waskul D, Gottschalk S and Ellis-Newstead T (2012) Making sense of the weather: dwelling and weathering on Canada's rain. Coast. *Space and Culture*, 15 (4): 361–380.

Wells GL and Loftus EF (2003) *Eyewitness memory for people and events*. In Goldstein AM (ed.) *Handbook of Psychology*. New York: John Wiley and Sons: 149–160.

Yusoff K and Gabrys J (2011) Climate change and the imagination. *Wiley Interdisciplinary Reviews: Climate Change*, 2 (4): 516–534.

2 Learning to say "Phew" instead of "Brrr"

Social and cultural change during the British summer of 1976

Ian Waites

Introduction

The British climate is generally accepted to be temperate and free from extremes on the whole (Rayner, 2009: 23). Nevertheless, it is a common view that the national identity of Britain is defined by an obsession with the weather and that the British love to talk about it on a daily basis (Harley, 2003: 103; Clifford and King, 2006: 435). This is fuelled by both a number of seemingly ancient sayings ("Rain before seven, fine before eleven") and conspicuously modern folklore tales. In April 2009, for instance, the Meteorological Office (hereafter Met Office) characterised their long-term weather forecast for that year with the promise of a 'barbecue summer' – which eventually turned out to be well and truly rained off (Harrabin, 2009). This also highlights one key question that is asked of Britain's weather every year: will it be a good summer or not? As the British naturalist Richard Mabey has put it, "Our climatic memory is ... full of gloomy mythology whose only silver lining is a vague belief in ancient Golden Summers" (2014: 21).

There are some places across the world – perhaps on the Algarve, or by the Mediterranean – where, every summer, you can generally expect to wake up day after day to blue skies and sunshine, almost without fail. The place where you might least expect this to happen is Britain. In 1976, however, the country experienced a summer heatwave that was notable not only for breaking all previous records for temperature, but also for its longevity: daytime temperatures across Britain climbed to at least 26.7°C (80F) every day between 22 June and 16 July, and this was a pattern that persisted until near the end of August (National Archives, 2009; Harley, 2015). As a consequence, the summer of 1976 has become *the* 'Golden Summer' in the history and mythology of British weather: in 2012, the *Daily Mail* reported on a survey on attitudes towards childhood, which suggested that the best time to be a child in Britain was during the summer of 1976.[1] Within one day over 300 reminiscences of that summer were posted on the newspaper's website. Comments such as this were typical:

> ohhhh i was 9 then......... omg how glorious. Off out as soon as my eyes opened and home in time for tea, which was usually egg on toast or

similar. jumping the brook, playing in the park, climbing trees, it was a brilliant time …never knew what my parents did, or how much money they had, back door was always unlocked … we didn't want anything else 'cos there were no commercials telling us how deprived we were if we didn't have the latest gadget… besides, we weren't in long enough to watch commercials. Tea, bath, bed, do it all again tomorrow.

(Mail Online, 2012. Accessed 20 June 2016)

The breathless, ecstatic shorthand of this comment brilliantly sums up both the immediacy of childhood and the sheer joy in remembering that summer. The adult perspective of the period was little different:

Wonderful days. I was not long married, we had very little in the way of material goods but we didn't care. The sun brought out the best in everyone – there was a sort of permanent holiday atmosphere around. My overriding memories are of a really terrific summer when we all felt life was pretty good and could even get better in the future. A great time to be alive.

(Mail Online, 2012. Accessed 20 June 2016)

Reminiscences like these give an impression almost of a paradisiacal dream world, to the point where they have also recently inspired a number of British novelists to use the summer of 1976 as a backdrop to their fictional explorations of social life and relationships in late-twentieth century Britain (Ashdown, 2013; O'Farrell, 2013; Cannon, 2016). Isobel Ashdown's *Summer of '76* for instance is set on the Isle of Wight and depicts the everyday becoming almost like something from a 1960s Californian beach movie, with nights warm enough for people to sit on the beach around driftwood fires, where the 'muted rhythm' of conversation 'rises and falls against the gentle lapping of the sea in the distance' (2013: 151).

But what was the reality of this summer? Between the happy memories and the fictional reconstructions of an event that is now 40 years past, what impact did it have on British society and culture at that time? What difference did nearly three months of continuously sunny, warm and, at times, very hot weather actually make to daily life in Britain, and how far did people embrace the heat – or make changes merely in order to cope with it? Scholarly research on the summer tends to be scientific in nature, variously dealing with the drought that year (Green, 1977), the effects of the heat on ozone levels, and the spread of algae; while the only study of the summer as a social or cultural phenomenon focuses exclusively on how British sport dealt with the unusual weather and its consequences (Kay, 2004). As a first study of its kind then, this chapter will draw upon a number of contemporary accounts of the summer as it unfolded: from local newspapers ("School sports days held in scorching hot sun"[2]) to the *Daily*

Mirror ("It's flaming June as Britain gets that Riviera touch"[3]); and from the diaries of Michael Palin to the letters of Kingsley Amis.

More critically however, it will evaluate the lasting impact of the summer by utilising those online reminiscences from the *Mail Online* website, together with others found on the online parenting forum *Netmums* (2012), the nostalgia website *Retrowow* (undated) and those collected in response to my requests for memories of the summer on a number of Lincolnshire community and local history group Facebook pages (2012). It has become commonplace now for many scholars, including geographers, to see memory as an expressive and actively binding force of social identity (Hoelscher and Alderman, 2004: 348). Furthermore, Edward Said has particularly noted that people reminisce, or 'refashion memory', in order to "give themselves a coherent identity, a national narrative, a place in the world" (2000: 179). The methodological problems in using Internet discussion groups and comments with regard to the British 'obsession' with the weather have been comprehensively discussed elsewhere (Harley, 2003) but, in general, the personal tone and content of the online reminiscences and comments seen here so far bear Said's notion out. They are also intensified by the recollection of the weather itself. While it is generally acknowledged that remembering the details of everyday life can be an uncertain process subject to what different people pay attention to and to how memories can be adulterated by subsequent, incorrect information (Fiedler et al., 1991; Wells and Loftus, 2003), studies vary on whether warm (Keller et al., 2005) or inclement (Forgas et al., 2009) weather has a positive effect on memory and mood. More of a consensus can be found in the remembering of high-profile weather events involving extreme temperatures or drought, which are often accompanied with emotional attachments or life reference points that are more likely to stick in the memory (de Vet, 2013: 199):

> What a summer. I picked up my brand new Kawasaki KE175 in green on my 17th birthday (April 8th) and had the most fantastic summer of my entire life. It seemed to run forever!! Day after day of endless sunshine, and life moving along at a snails pace. Long, balmy summer evenings without a breath of wind to cool you down. Out on my bike till all hours just loving the feeling of freedom, and independence that it gave to me. I could go anywhere whenever I wanted to!!
>
> (Retrowow, undated. Accessed 20 June 2016)

Or:

> I was 17 and living in Cardiff during that long hot summer, I had my first Motorbike, a Suzuki TS185 I loved that bike. Taught my first girlfriend to ride it, she was so beautiful I was madly in love with her, her name was Julie. There was a group of us all about the same age most had

motorbikes and we would spend the summer going for rides, going to the pub or just sitting in the field where we used to meet up, talking and messing around till it was late. A small stream ran along the edge of the field, this had dried to just a trickle, the smell of scorched earth bracken and grass filled the air. I used to take a radio to the field on a Sunday evening and we would listen to the top 20 songs.

(Retrowow, undated. Accessed 20 June 2016)

On one hand, these reminiscences might collectively come across as somewhat generic in nature – dreamy, a little over-romanticised, and characterised by soft-focussed, nostalgic turns of phrase. On the other, there are marked specificities here, both in the popular cultural detail and in the way that these sentiments are sincerely and *keenly* expressed:

So many great postings here, all of which I can relate to. Happy, happy times. I was 20 in April that year ... It bugs me when I hear remarks about the 70s such as 'the decade that taste forgot.' Obviously made by people who never experienced the halcyon days when life was slower and people seemed to have so much more time for each other. Memories of happy times when Dad was still with us, and the whole family were together ... Whatever happens in the future, at least I have the memories of endless days with blue skies, and people happy, all around me. No-one can take that away.

(Retrowow, undated. Accessed 20 June 2016)

'Without known precedent': the weather during the summer of 1976

'Endless days with blue skies' seemed unlikely at the beginning of June 1976, when the *Times* published a Met Office weather forecast for the coming month. "More Rain Likely" went the subheading as "rather changeable weather" was set to "predominate over the month as a whole". A "few brief warm spells" were possible but temperatures and sunshine levels were both expected to be "below average in most places".[4] This proved correct up to the start of the 90th Wimbledon tennis championships on Monday 21 June, when overcast skies and mid-sixties temperatures were recorded. The next day however signalled a change – all of a sudden, clear sunshine shone over London, and temperatures reached nearly 24°C (75F). Three days later, the *Times* stated that it was "85 degrees in the shade on the centre court", where the heat was so oppressive that people were fainting in numbers. "They're rolling in all the time" said a man at the first-aid post with what seemed like an inappropriate cheerfulness, "We're nearing a record".[5] At the start of the following week, the *Daily Mirror* reported that the temperature "in loopy London" was a "gasping 93 degrees".[6] By then the Wimbledon fortnight had come to a close, and a newcomer, "Bjorn Borg, 20" had won the men's championship (beating Ilie Năstase 6–4 6–2 9–7), leaving the *Times* to sum it all up as "an overheated tournament" dominated by "two icebergs glistening in the heat".[7]

By the end of June we were all 'glistening in the heat'. The *Times* reported that the Met Office long-range forecasting team was already admitting that the summer was turning out to be "without known precedent". The article went on to note that the highest ever June temperature at that time – 35.6°C (96.08F) – had already been recorded at Mayflower Park in Southampton on Monday 28 June, and even the *Times* glistened a little under the influence of the spectacular summer weather: "The recent hot, dry weather seems likely to continue for several days in most districts and over the month as a whole a good deal of warm, mostly dry weather is expected". Indeed, the article concluded by providing some uncharacteristic pieces of so-called 'silly season' summer trivia: at Dudley Zoo, special sunshades had to be obtained for the penguins.[8]

They were going to need them: on Saturday 3 July the temperature hit a record-breaking 35.9°C (96.6F) in Cheltenham, just one of 11 consecutive days of temperatures over 31°C (89F) there (National Archives, 2009). It is worth noting at this stage that recorded temperatures are of the air surrounding the thermometer, which must always be shaded from sunlight and exposed to adequate ventilation – Cheltenham was therefore experiencing daily temperatures of up to 36°C *in the shade*. "WHEN will the great super-sizzle end?" asked the *Daily Mirror* during the first week of July.[9] It was a good question – from 23 June to 8 July for instance, Heathrow Airport had recorded 16 consecutive days with temperatures over 30°C (86F), its longest spell of hot weather on record. The research for this chapter partly involved poring over the daily weather forecasts from both the *Times* and the *Daily Mirror* for June, July and August of 1976, and it soon became clear from this that the answer to the *Mirror's* question was exactly the same as the one it gave back then: "NOT BY A LONG SWEAT". Give or take an odd seven days or so of thundery weather, the newspaper forecasts revealed a situation that more or less added up to *two months* of continuously sunny, very warm and – at times – very hot weather. Day after day, the *Daily Mirror's* forecast read "Dry, sunny, very hot". The outlook was always "Similar".[10] This also comes through in the reminiscences of the time:

> We lived on the south coast of England and were never indoors…we were brown as berries and felt so healthy and energetic … Day after day started misty and soon the mist burned off, and then it was baking!
>
> (Mail Online, 2012. Accessed 20 June 2016)

And again:

> The summer of 76, what can you say. We ended up taking it for granted that tomorrow was gonna be hot (which it was). I was 19 at the time

and working shifts in a local factory. On nights we used to have our last break at 4 till 4.30 and go outside to watch the sun come up.

(Retrowow, undated. Accessed 20 June 2016)

Said's notion that individual reminiscences can form part of a 'national narrative' is supported here by the fact that the heatwave was by no means confined to the South: it was happening everywhere, from the Channel Islands up to Wales, across all of the industrial Midlands and North, and beyond. On Sunday 4 July, Blackpool enjoyed a maximum temperature of 31°C (88F), seven degrees higher than Bournemouth on that day.[11] One on-line commentator, from Redcar on the North East coast, remembered the summer as the only time "the North Sea actually felt warm to go swimming in" (Retrowow, undated. Accessed 20 June 2016). Even Fair Isle, the most remote and northern inhabited community in the British Isles, was experiencing daily temperatures throughout late June and early July in the low to mid-seventies Fahrenheit.[12]

And so it went on: through July and into August, a month when daily temperatures across England and Wales averaged between 24°C and 26°C (75–79F). Between 16 and 28 August, the temperature in many places once more nudged into the eighties Fahrenheit (National Archives 2009). At the end of the summer, it was noted that a weather station on the West Common at Lincoln had recorded an extraordinary 74 consecutive days of sunshine, from 17 June to 30 August.[13] Nationally, the heatwave provoked a decline in consumer spending.[14] Perhaps understandably, people were staying away from the overheated, stuffy shops (very few had air-conditioning back then), but this was also an indication of how they were beginning to modify their usual patterns of everyday life in order to adapt and mitigate the impact of the prolonged heat. The fall in consumer activity was one factor in this, but it was also indicative of two more dominant beliefs that come through repeatedly in the reminiscences of the summer: people were perhaps less concerned with material goods back then, and that a 'permanent holiday atmosphere' began to set in as the sunshine caused people to spend more time outdoors with each other.

'Relaxing in the warmth and getting to know each other': changes in attitude, behaviour and lifestyle

One memory of the summer in Lincoln was not about emerging into hot, sunny mornings after a night shift, or of picking up girls on a brand new Kawasaki KE175. Instead, the summer was significant because the correspondent recalled that it was the first year that he kept honeybees. The weather was very dry and many plants withered, but the summer also provided 'huge crops of delicious clover honey'. In many ways, everyday life during the summer followed some very traditional patterns made up of small pleasures and formed by a sense of quiet respectability – indeed,

I was offered 'a full record' of the correspondent's honey production for that year, with dates, if it was of any use to me (Facebook, 2012. Accessed 6 July 2014). Interesting as that kind of detail might have been, I turned down the offer with thanks; this particular memory of the summer nevertheless presents to us a world far removed from the more superficial images of the 1970s.

Other, equally modest pleasures of everyday life were boosted by the heat elsewhere in Lincolnshire, in the market town of Gainsborough. Whites Wood Junior School's 22nd annual sports day was held during the first week of July in "scorching hot sun".[15] Over the weekend of 9–11 July, the heat drew record crowds to the Morton Feast just outside Gainsborough, where a "record sum" was spent, and when the new 'Miss Morton 1976', 17 year-old Gainsborough High School girl Katrina Watson, was crowned. She received 'a sash', a bouquet of flowers and a voucher for 'a shampoo and set' by local hairdressers 'Studio One'. Various children's races took place where, in the egalitarian spirit of this still very much post-war society, "each child that entered a race was given a bag of sweets". There was a best-kept garden competition and a 'baby competition'. Eleven-year old Shannon Walker won the over-sevens category of the children's fancy-dress competition dressed topically (and presciently as the summer went on) as 'Water Shortage', in a 'cardboard stand-pipe dress'.[16]

Lest we begin to view such activities today as being quaint, suburban, naïve even, they nevertheless continue to deny dominant preconceptions of the brash and tasteless 1970s, tending instead towards a number of recent social-historical studies of community in post-war Britain (Tebbutt, 2012; Watkiss Singleton, 2011) that have noted how mid-century patterns of restraint, neighbourliness and respectability in everyday life persisted some time into the decade. We should also consider the wider social and economic situation as it was in 1976, compared to how it has become in recent years. In January 2010, a government-commissioned report, *An Anatomy of Economic Inequality in the UK*, demonstrated that, despite outward impressions of affluence and comfort, the country had become more unequal than ever before. Moreover, the report stated that between 1937 and 2010, 1976 was the year when the inequality gap was at its narrowest: 'only' 4.17% of the UK's total income was going to the richest 1%.[17] A 2007 letter to the *Guardian* provided a wider social and cultural preamble to this particular finding, by railing against the popular notion that things were especially grim in the 1970s:

> For the majority of British people, they were the best times ever. Housing was affordable, students had grants, eye and dental check-ups were free at source, car parking was usually free, the work environment was more relaxed, working hours were shorter, the retirement age was lower, it was far cheaper to watch sports events, the service provided by the public utilities was far superior, and so on.

"Somebody", the correspondent concluded, "needs to ... nail the ludicrous myth that we are better off now".[18]

A more detailed analysis of the socioeconomic context of the 1976 summer largely falls outside the remit of this chapter, but much of what this letter stated was undoubtedly true – for instance, it cost 70p to see Manchester United play Leeds United in the pre-Premier League, League Division One in March 1976, still only around £4.00 in today's values – and it gives a more objective impression at least of a settled, more open and less demanding time.[19] This again comes through repeatedly in the reminiscences of the summer:

> Yes, life did move at a much slower pace. No-one bumped into you on the street because they were jabbering into a phone or texting, and there wasn't the email and technology-induced stress we have now.
>
> (Retrowow, undated. Accessed 20 June 2016)

> There's no doubt life was less stressful for adults and kids then. We had no money but we didn't have the high expectations that people have now. We didn't want so much. Kids were happy playing out, camping in the garden ... Most dads got home from work at 5 o clock-ish-my hubby gets in at 7 now.
>
> (Netmums, 2012. Accessed 13 June 2016)

> Never had a summer like it since, and in some ways even if we did it wouldn't be the same. The pace of life is different now. Expectations are too high ...
>
> (Retrowow, undated. Accessed 20 June 2016)

The slower pace would have certainly been encouraged by the heatwave, but the most significant social transformation that occurred as a direct consequence of the weather was that people began to spend more time socialising outdoors on a day-to-day basis. Forty years on, the evidence for this can only be anecdotal, but 60% of the reminiscences researched here recall a summer when life was predominantly spent outdoors, which roughly coincides with a recent study of the UK public's recollections of unusually hot and dry summers in the 1990s, when 66% of respondents also cited increased outdoor activity as a positive impact on their everyday lives (Palutikof et al., 2004: 48–49). The locus for this was the domestic garden. Surveys throughout the second half of the 1970s consistently paint a traditional picture of the home being at the centre of people's lives where, for instance, over half of all men, and up to a quarter of women spent time gardening (Sandbrook, 2012: 15). During the summer of 1976, however, the garden became less of an ornamental expression of private aspiration and more of an extension of the home itself as a new mode of outdoor socialising attempted to 'domesticate' the effects of the persistent heat (Rayner, 2009: 19). On Saturday, 2 July 1976, Michael Palin wrote in his diary that he was:

> determined to drink only moderately today. Failed hopelessly. No sooner had last night's mixture of lager and cider seeped through my system than it was quietly being worked on by some very more-ish sangria at the Denselows.

Palin was there for a lunch party of "media friends from Panorama, Bush House, The Guardian etc.", but this "turned into a very large barbecue ... out in the blazing heat" (2007: 360). This sort of thing was not restricted to the media bourgeoisie of Gospel Oak, however. In Northern Ireland, Maggie O'Farrell recalled that "Our parents and their friends sat out on their patios until late in the evening, radios threading out through the opened windows" (2013: 332). Equally, on a new council estate in Bridgend, the weather helped to forge a sense of community:

> July 1976 was when we moved into a new house ... every evening we and all the neighbours sat on the front lawn or doorstep having tea, coffee or booze, up to about 10:30 every evening until the autumn set in, relaxing in the warmth and getting to know each other as we were all new to the street.
>
> (Retrowow, undated. Accessed 20 June 2016)

This new aspect of outdoor socialising easily extended well into the evening, helped by average night-time temperatures of at least 17°C (63F). Children camped out overnight: "I was 7 years old at the time and remember camping out in the back garden for most of the summer as it was so hot" (Mail Online, 2012. Accessed 20 June 2016). Adults too: "some of our neighbours used to sleep outside in their gardens because it was so warm at night" (Retrowow, undated. Accessed 20 June 2016).

For a few fleeting months, people were able to spend more time living what one reminiscence described as "an outdoor lifestyle that was distinctly and deliciously un-British" (Netmums, 2012. Accessed 13 June 2016). Even pub-goers were venturing outside and away from the dark, Victorianate 'snugs' and 'lounge bars', instigating the rise of the beer garden and the promise of a more sophisticated and less insular mode of public drinking. A "snatched lunchtime drink in a Campden Hill pub" suddenly "felt like a holiday" as everyone began to take drinks outside onto the pavements and into the sun: "Sunshades went up. Soon it looked more like Madrid or Rome than Manchester or London" (Speller, 2006: 151). For a nation generally resigned to mediocre summers, the consolations of hunkering down in the corner of the local were considerable, and the idea of taking your drink outside of the pub at that time was almost unknown: in London's West End, the actor Kenneth Williams was shocked to see "everyone standing *outside* pubs holding beer in their hands" (Williams, 1994: 521). But this summer was different and presented a balmy world of new, seemingly limitless and evermore sophisticated social pleasures: "This was the summer that got us all sitting outside at pubs, having BBQs in the back gardens, things that are the norm now" (Retrowow, undated. Accessed 20 June 2016).

Drinking obviously fuelled the beginning of Kingsley Amis' letter to his friend, the poet Robert Conquest, on 5 July: "Dear Bob, Fucking amazing weather continues." Amis was in the process of moving house, and his "main job" was to "drink up the nearly-empty bottles" of "horrible stuff, like cherry vodka, Mavrodaphne, raki etc.", which clearly demonstrated the wide-ranging tastes of the well-travelled educated classes at this time, and Amis would have spoken with close knowledge when he said that it was 90° in London and therefore "hotter than Sicily" (Amis, 2001: 800). At the same time, such 'foreign' tastes were being cultivated across the social spectrum due to the rise of the foreign package holiday: at the end of July, an article in the *Gainsborough News* reported that the number of passports issued had topped the million mark for the first time. Back then, a passport cost £4 and, in the year up to the end of June, a total of 1,477 had been issued in the Gainsborough area alone.[20]

The boom in foreign travel was temporarily halted by the guaranteed heat and sunshine, which was persuading people to stay at home: according to the *Gainsborough News*, the "usual exodus" at the start of the summer holidays had been "stemmed" and "holidaymakers are taking advantage of the numerous day trips being run in the town". The summer also appeared to be ushering in a decline in traditional work and leisure patterns, and the beginnings of a shift towards a more contemporary flexibility. Since the nineteenth century, factory workers took their annual summer holiday in August during 'Trips', or 'factory fortnight', but one local tour operator in Gainsborough pointed out that "trip holidays are not as rigid as they used to be with a lot of firms now staggering their holidays".[21] Day trips to seaside resorts were thriving instead with "two or three coaches going out full every day", a situation that again coincides with the findings of Palutikof et al., that hot summers in Britain increase the likelihood of people going on more day trips, as opposed to planned, extended holidays (2004: 52).

The *Daily Mirror* summed all this up with a cartoon of a man saying to a couple: "Gosh, you two look pale. Did you go abroad for your holiday?"[22] The idea that Britain was undergoing a number of social and cultural transformations due to the heatwave was beginning to be noted. Indeed, the heat was literally changing tastes: the traditional English bacon-and-egg breakfast was becoming a "mouth-watering memory for thousands of holiday-makers" as the Pontins chain of holiday camps was finding that the summer was cutting demand for fry-ups, forcing the company to switch to a "Continental-style rolls and butter breakfast".[23] A week after the *Mirror* had headlined with "It's flaming June as Britain gets that Riviera touch",[24] its Comment section was suggesting that the heatwave was "knocking the stuffing out of some hallowed traditions", starting with the news that for the first time in its 189 year history, the Marylebone Cricket Club had allowed members to take off their jackets at Lord's cricket ground. "Cherished notions about the British character", the *Mirror* went on, were "biting the dust" and "With a bit more sun", we might become more like "those excitable Europeans". It was suggested that we should finally drop Fahrenheit for Centigrade and "learn to say 'Phew' instead of 'Brrr' when the forecast is 32°". Another point – that we should start to take 'sensible siestas' – was already taking hold as some factories were enabling their workers to change their shift patterns and breaks to avoid the stifling heat at work and to rest if they needed to. Britain's new outdoor drinking habits also prompted the notion that licensing laws should be relaxed in a more European manner, so that it was possible to "get a refreshing pint at four in the afternoon when it's needed the most".[25]

Despite these proposed opportunities for cultural enlightenment, the tabloid understanding of 'those excitable Europeans' was nevertheless somewhat misconceived. As temperatures persisted in the 30°C over the first week of July, it was reported that a Frenchman who was arrested for bathing nude in the Trafalgar Square fountain told police "I couldn't do this in France".[26] Dressing down to some degree at least seemed to be a necessity if you were to survive the extreme temperatures. This was not a problem if you were a child: "I don't think I ever got dressed that summer. Aged four, what was the point?" (O'Farrell, 2013: 331). For an adult in 1970s Britain however, it was still a question of limits. A 48-year-old man was suspended without pay at a factory in Slough because he turned up for work wearing shorts. He was reinstated the next day but was told that he could wear them only as long as the heatwave lasted.[27] A few days later, the *Mirror* published a letter from a Mrs Heath of Farnham, Surrey:

> During the hot spell lots of people have walked topless in the streets and not an eyebrow has been raised. Why? Because they are men. But if a woman presumes to do likewise she risks finding herself in court. Setting aside the relative attractions of the upper female form as opposed to that of the male, why this discrimination in these days of much vaunted equality?

Any answer to what was, in itself, an inherently compromised point of view remained evasive in this period of casual sexism, especially when this letter was published alongside a feature entitled "The prettiest girls are always in the Mirror" with a topless photograph of "Maria", a "student who designs clothes" but who "hasn't been wearing a lot lately because of the hot weather".[28]

Advances in remedying the deep-rooted marginalisation of women at this time were not going to be made as a consequence of the heatwave, but the exceptional weather did entice people outdoors and encouraged them to improvise with traditional and everyday social and spatial realms in a manner that would soon come to be characterised as 'lifestyle' activities. People did slow down, dress down and take siestas, and they sat out drinking and socialising with friends and neighbours at night. At the same time though, the extreme weather left people in wonderment at the fact that washing could dry on the line almost "in seconds, stiff and crisp as if frozen" (Facebook, 2012. Accessed 6 July 2014) but also fearful that the grass everywhere was being "burned to a crisp" (Mail Online, 2012. Accessed 20 June 2016). Another, more perceptive comment stated that "Other contributors have heralded that summer as a sociological turning point ... I wonder if that was true. I don't think I look back with rose tinted glasses ... most of the roses had died through lack of water" (Retrowow, undated. Accessed 20 June 2016).

"An implacable enemy": the summer of 1976 as an extreme weather event

The 1976 heatwave was already riding on the back of the driest 16-month period in 250 years, and only half of the normal rainfall usually expected over a typical British summer fell in June and August that year (Green, 1977). Maggie O'Farrell stated that when she started to research her summer of 1976 novel *Instructions for a Heatwave*, she soon "realised the severity, the seriousness of what happened. We all talk about the 'heatwave' of 76 but the more correct term would be 'drought'". To her, the glorious summer weather became "an implacable enemy" (2013: 331–332). Debatably, the real 'enemy' was society itself. The water shortages brought on by the drought were nevertheless exacerbated by the ways in which people were responding to the heat. Water consumption increased dramatically in the first few days of the heatwave: the Thames Water Authority reported that Londoners were consuming up to 600 million gallons of water per day and that the supply infrastructure was struggling to cope with the demand. In response, the Water Authority called for each household to reduce its consumption by five gallons a day, a saving that could be achieved simply by running a garden hose for 45 minutes instead of an hour.[29]

By mid-August however, Britain's ten water authorities were seeking to extend bans to the domestic use of hosepipes because, according to one authority spokesman, "selfish" people "didn't care about saving water".[30] Many reminiscences of the summer support this, with references for instance to "sneaky neighbours running sprinklers at night" (Mail Online, 2012. Accessed 20 June 2016). The newly found enthusiasm for garden socialising was taking its toll: "We'd have our own stupid its-a-knockout competitions in the back garden, with hoses gushing, not realising what a hosepipe ban was" (Retrowow, undated. Accessed 20 June 2016). Bleaker still was this single childhood memory: "Dead grass and my dad chasing me with the hosepipe" (Facebook, 2012. Accessed 6 July 2014). As a consequence, some water authorities in the South and South West were forced to ration domestic consumption through the use of standpipes, as arable and vegetable crops were failing amongst talk of rising prices and food shortages.[31]

As such, the summer of 1976 can certainly be defined as an 'extreme weather event', in that it was represented by extreme values that were likely to cause damage such as critical water shortages and crop failure (Stephenson, 2008: 12–14). This was shockingly visible: the *Daily Mirror* published an undoubtedly bleak photograph of a teenage girl placing her hand down into the fissures of the cracked and baked Pitsford Reservoir in Northamptonshire, where the water level had dropped 20 ft.[32] One memory of seeing England from an aeroplane returning from abroad went like this:

> the landscape which unrolled beneath … was an alien one. Earth-brown fields, silvery crops, limp-leafed hedges and trees which seemed to be already touched with autumn, spread across the Sussex Downs. We passed over a reservoir, its pale clay sides containing just a dark sump of water and in the last miles smoke rose from blackened woodland into the clear sky. When I stepped out of the plane, the tarmac was soft beneath my feet and the sparse grass between the runways was deader than where I had come from, a thousand miles south.
>
> (Speller, 2006: 151. With kind permission from Elizabeth Speller/Vogue © The Conde Nast Publications Ltd.)

In this sense, the drought was indeed becoming an implacable enemy, as one *Times* article alone indicates:

> Huge fires were sweeping parts of Britain last night, putting a severe strain on fire services and threatening property. Thousands of acres of wood and heathland were ablaze in Surrey. One of the biggest fires covered more that five square miles of countryside ranging across Elstead and Thursley.

At Thursley, bulldozers had to be brought in to dig trenches around the village to protect homes. At Whipps Cross Lido, by Waltham Forest in London, thousands of swimmers had to be led to safety when a forest fire surrounded the pool. On the North York Moors, it took fire fighters four days to contain a fire that destroyed 50 acres of trees.[33]

The newly found enthusiasm for outdoor living in general was having further adverse consequences. Police warned that Laughton Forest, near Gainsborough, was in danger of becoming a fire trap, following a number of fires that had been caused by picnickers dropping cigarettes and abandoning camp fires.[34] The warnings fell on deaf ears: a month later the Lincolnshire County Council Fire Chief 'slammed' the public for their "total apathy and irresponsibility" after a spate of similar fires. The county fire service had reached 'crisis point' after one day alone saw 60 calls being made, where all but four were concerning grass fires.[35] Nationally, the Forestry Commission and the National Trust attempted to apply smoking bans to their land, and teams of volunteers formed fire-watch groups. The overall scene was almost apocalyptic in manner: in Dorset, peat beds spontaneously burst into flames, and in Cornwall fires spread with flames seen travelling across land at 40 miles an hour.[36] A letter to the *Gainsborough News* described how "This morning a black cloud passed over our village and discharged a copious shower of burned and burning straw all over everyone's property" as farmers continued to burn harvest stubble despite requests from the National Farmers Union to defer this practice while the hot and dry weather continued.[37]

An even more crucial, but hitherto largely unconsidered, context to this recklessness was raised in a *Times* article that appeared at the end of August. Looking back, the piece argued that the weather was no longer a "matter of a long, dry summer in Britain"; rather, it was a "global problem" affected by "man-made influences". Recent scholarship (Hulme, 2014) has stressed the many difficulties in attributing extreme weather events to climate change, but a related general discourse was clearly emerging in 1976 when, according to that *Times* article, the circumpolar jet stream had "zig-zagged", bringing not only extreme heat to Britain but also daily deluges to Hong Kong and Moscow, all of which, the article went on, "attests to the fact that the world's climate generally is out of joint".[38]

Conclusion: 'More like the Britain we know'

In 1976, 'climate change' was not yet a topic of widespread public knowledge or discussion. Even today, global warming and climate change are not seen as pressing risks, with more concern being directed towards more immediate concerns about the everyday effects of pollution (Guber and Bosso, 2012: 59). It is only to be expected therefore that the media in 1976 would place so much stress on the direct and increasingly deleterious environmental consequences of the drought. It is equally clear that the general public, both at the time and from the perspective of reminiscences today, nevertheless relished the heatwave. Enjoying the prospect of guaranteed heat and sun, worrying about some of its consequences and – indeed – disregarding them were not mutually exclusive. Given Britain's temperate climate, and a national identity that is sometimes expressed by a general expectation of disappointing summers, it is perhaps understandable that the British have been found to prefer unusually warm weather to unusually cold weather (Palutnikof et al., 2012: 48) – a situation that is compellingly borne out by the positive and sometimes ecstatic reminiscences of the 1976 heatwave.

The *Times* piece alone demonstrates just how the 1970s were more significant in the shaping of the twenty-first-century world than we might at first imagine. In 1970s Britain, a combination of greater social autonomy and increased affluence gave rise to the more commodified and aspirational behaviour of the 1980s and beyond. However, if the 'swinging sixties' sparked certain social and economic freedoms, and the 1980s brought a consequential sense of aspiration and individualism to an extreme, the 1970s might therefore be understood as a pivotal decade that provided a balance, a check even, between the two extremes on either side. In a number of subtle but pervasive ways, the summer of 1976 fuelled this, allowing traditional ways of life to flourish, while opening new social horizons.

When we look back over the last 40 years, from today's fragmented social and economic climate to recall that summer, we could be forgiven for thinking that it was not just the best time ever; it was the *final* best time ever. For those of us who were children or teenagers then, it is embedded in our psyche. Arguably, memories of that childhood six-week summer school holiday might always appear to be sun-filled only because we were out playing all day, every day. It could also be supposed that the memories of endless sunshine, of hot days and warm, sticky nights were actually based only on the blurring of several, intermittently exceptional, days across the whole of that summer: according to Harley (2003: 115) people tend to be nostalgic about past weather events when they have few statistical grounds for doing so.

But the statistics for the summer of 1976 show that its remembrance was no trick of memory – the summer went on and on, and it therefore represents what was probably the only truly national, extended and commonly experienced period of warmth (in all senses of the word), serenity and liberty in the history of twentieth-century Britain. It was also destined not to last. The weather for the week commencing 23 August again predicted more "Dry, Sunny" and either "Very Warm" or "Hot" days with maximum temperatures across most of the British Isles of 27°C (80F).[39] The August Bank Holiday loomed – surely a hot and sunny long weekend would come to crown this summer of all summers, but it was not to be. On Bank Holiday Monday, the front page of *The Times* announced: "Rain floods homes but not the reservoirs": homes in the South West were flooded, motorway drivers were "forced to a halt", and day-trippers abandoned their outings as thunderstorms and heavy showers "swept many parts of Britain".[40] The *Daily Mirror* that weekend displayed a front-page photograph of a man with an umbrella and asked "Know what this is, folks? ... In Britain yesterday there were umbrellas galore. Thousands *unfurled* in London with rust squeaking groans. For yesterday, it RAINED". In its almost gleeful tabloid manner of wringing every last amused drop out of the occasion, the paper went on to say that showers "spattered" the country, and windscreen wipers "creaked into action". And that, it concluded, was "more like the Britain we know".[41]

"Where else in the world would the weather wait two months to rain on a bank holiday weekend?" lamented the same newspaper a few days later.[42] As if right on cue, the national mood suddenly darkened: during the final days of the school summer holidays, a "changeable N to NW airstream" was covering the British Isles.[43] As schoolchildren were preparing to go back to school, there was a resurgence of the kind of industrial unrest that was to pave the way to the social and economic winter that defined the final years of the decade. "ROAD TO RUIN", the *Daily Mirror* cried, as 33,000 British Leyland workers were threatening to be on strike by the beginning of the new working week.[44] No more "Dry, sunny and very hot: Outlook Similar". The carefree days of the 1976 summer heatwave had undoubtedly come to an end, and normal British life resumed: "when the rain fell everyone came out of their houses and just stood there ... such was the relief. Of course within two weeks we were back to complaining about it" (Retrowow, undated. Accessed 20 June 2016).

Notes

1 Reilly J. and Brooke C. 'Dancing Queen, Raleigh Choppers and space hoppers: How 1976 was the best summer to be a child in Britain'. *Daily Mail*, 19 March 2012.
2 *Gainsborough News*, 'Sports Day Roundup'. 2 July 1976.
3 Hebditch, R. 'It's flaming June as Britain gets that Riviera Touch'. *Daily Mirror*, 24 June 1976, p. 2.
4 *The Times*, '30-day forecast'. 2 June 1976, p. 2.
5 Bellamy, R. 'Britain's youngsters can look back with pride'. *The Times*, 24 June 1976, p. 11.
6 Todd, R. '93° The Big Heat'. *Daily Mirror*, 29 June 1976, p. 5.
7 Bellamy, R. 'Two icebergs who dominated a half-cooked competition'. *The Times*, 5 July 1976, p. 5.
8 Hodgkinson, N. 'Hottest June of century surprised the meteorologists'. *The Times*, 29 June 1976, p. 2.
9 Bedford, R. 'Scorched Earth'. *Daily Mirror*, 6 July 1976, p. 5.
10 *Daily Mirror*, 'Weather World'. 7 July 1976, p. 2.
11 *Daily Mirror*, 'Weather World'. 5 July 1976, p. 2.
12 *The Times*, 'Weather forecast and recordings', 28–30 June; 2–6 July 1976, p. 2.
13 *Lincolnshire Chronicle*, 'Record summer outshines the rest'. 23 September 1976.
14 Emler, R. 'Consumer spending stays sluggish'. *The Times*, 20 August 1976, p. 16.
15 *Gainsborough News*, 'Sports Day Roundup'. 2 July 1976.
16 Cowley, G. 'Record 'gate' spends record sum at village feast'. *Gainsborough News*, 16 July 1976.
17 Gentleman, A. and Mulholland, H. 'Unequal Britain: richest 10% are now 100 times better off than the poorest'. *The Guardian*, 27 January 2010. Available from www.theguardian.com/society/2010/jan/27/unequal-britain-report (accessed 20 June 2016).
18 *The Guardian*, 'Letters to the Editor: In defence of the 70s'. 13 October 2007.
19 Campbell, P. 'When football was reasonably priced: your old ticket stubs from years gone by'. *The Guardian*, 17 February 2015. Available from www.theguardian.com/football/blog/gallery/2015/feb/17/football-ticket-prices-old-ticket-stubs (accessed 20 June 2016).
20 *Gainsborough News*, 'Passport boom in Gainsborough area'. 30 July 1976.
21 *Gainsborough News*, 'Days out take over in Trips fortnight'. 30 July 1976.
22 *Daily Mirror*, 7 July 1976, p. 11
23 White, R. 'Bacon 'n' egg crunch'. *Daily Mirror*, 9 July 1976, p. 3.
24 See note 3.
25 *Daily Mirror*, 'Mirror Comment: 32° Phew'. 30 June 1976, p. 2.
26 *Daily Mirror*, 'Hot News Extra'. 6 July 1976, p. 5.

27 *Daily Mirror*, 'Flare-up over Mr Cool-Legs'. 2 July 1976, p. 3.
28 *Daily Mirror*, Public Opinion'. 14 July 1976, p. 4.
29 McCarthy, M. 'Turn 'Em Off''. *Daily Mirror*, 2 July 1976, p. 3
30 Plaice, E. and Davies, M. 'Fire! It's the water bomber'. *Daily Mirror*, 17 August 1976, p. 2.
31 *Daily Mirror*, 'Food and £ wilt in sun'. 25 August 1976, p. 5.
32 Bedford, R. 'Scorched Earth'. *Daily Mirror*, 6 July 1976, p. 5.
33 *The Times*, 'Moorland ablaze'. 1 July 1976, p. 1.
34 *Gainsborough News*, 'We're having a heatwave but phenomenal fires increase'. 2 July 1976.
35 *Gainsborough News*, 'Fire Chief slams "firebugs"'. 6 August 1976.
36 Riley, J. 'The record summer of 1976'. *Liverpool Echo*, 16 August 2004. Available from http://www.liverpoolecho.co.uk/news/liverpool-news/the-record-summer-of-1976-3540938 (accessed 20 June 2016).
37 *Gainsborough News*, 'Why do they burn?' 27 August 1976.
38 Spanier, D. 'Talking about the weather must be more than a national pastime'. *The Times*, 28 August 1976, p. 10.
39 *The Times*, 'Weather forecast and recordings'. 23–26 August 1976, p. 2.
40 *The Times*, 'Rain flood homes but not the reservoirs'. 30 August 1976, p. 1.
41 *Daily Mirror*, 28 August 1976, p. 1.
42 *Daily Mirror*, 'Mirror Comment: Just like old times'. 30 August 1976, p. 2.
43 *The Times*, 'Weather forecast and recordings'. 1 September 1976, p. 2.
44 Connew, P. and Stringer, T. 'On the Road to Ruin'. *Daily Mirror*, 4 September 1976, p. 1.

References

Amis K (2001) *The Letters of Kingsley Amis*. London: Harper Collins.

Ashdown I (2013) *Summer of '76*. Brighton: Myriad Editions.

Cannon J (2016) *The Trouble with Goats and Sheep*. London: The Borough Press.

Clifford S and King A (2006) *England in Particular*. London: Hodder and Stoughton.

de Vet E (2013) Exploring weather-related experiences and practices: examining methodological approaches. *Area*, 45 (2): 198–206.

Facebook (2012) Available from www.facebook.com/groups/ypfli/ (accessed 6 July 2014).

Fiedler K, Asbeck J and Nickel S (1991) Mood and constructive memory effects on social judgment. *Cognition and Emotion*, 5: 363–378.

Forgas JP, Goldenberg L and Unkelbach C (2009) Can bad weather improve your memory? An unobtrusive field study of natural mood effects on real-life memory. *Journal of Social Psychology*, 45 (1): 254–257.

Green JSA (1977) The Weather during July 1976: some dynamical Considerations of the drought. *Weather*, 32: 120–126.

Guber DL and Bosso CJ (2011) High hopes and bitter disappointment: public discourse and the limits of the environmental movement in climate change discourse. In Vig NJ and Kraft ME (eds.) *Environmental Policy: New Directions for the Twenty-First Century*. Washington: CQ Press: 54–82.

Harley TA (2003) Nice weather for the time of year: the British obsession with the weather. In: Strauss S. and Orlove BS (eds.) *Weather, Climate, Culture*. London: Berg: 103–118.

Harley TA (2015) The hottest and coldest summers and winters. www.trevorharley.com/trevorharley/weather_web_pages/hot_summers_cold_winters.htm (accessed 5 August 2016).

Harrabin R (2009) Met office cools summer forecast. London: BBC. Available from http://news.bbc.co.uk/1/hi/8173533.stm (accessed 20 June 2016).

Hoelscher S and Alderman DH (2004) Memory and place: geographies of a critical relationship. *Social and Cultural Geography*, 5: 347–355.

Hulme M (2014) Attributing weather extremes to 'climate change' a review. *Progress in Physical Geography*, 38 (4): 499–511.

Kay J (2004) Dust to dust: The summer of 1976. *Weather*, 59 (9): 247–250.

Keller MC, Fredrickson BL, Ybarra O, Côté S, Johnson K, Mikels J, Conway A and Wager T (2005) A warm heart and a clear head: the contingent effects of weather on mood and cognition. *Psychological Science*, 16 (9): 724–731.

Mabey R (2014) *A Brush with Nature: Reflections on the Natural World*. London: Ebury Press.

Mail Online (2012) Available from www.dailymail.co.uk/news/article-2117115/Dancing-Queen-Raleigh-Choppers-space-hoppers-How-1976-best-summer-child-Britain.html#comments (accessed 20 June 2016).

National Archives (2009) Great Weather Events. Available from http://webarchive.nationalarchives.gov.uk/+/http:/www.metoffice.gov.uk/corporate/pressoffice/anniversary/summer1976.html (accessed 20 June 2016).

Netmums (2012) The happiest summer for children – was it 1976? Available from www.netmums.com/coffeehouse/general-coffeehouse-chat-514/news-current-affairs-topical-discussion-12/737721-happiest-summer-children-1976-a.html (accessed 13 June 2016).

O'Farrell M (2013) *Instructions on a Heatwave*. London: Tinder Press.

Palin M. (2007) *Diaries 1969–1979: The Python Years*. London: Phoenix.

Palutikof JP, Agnew MD and Hoard MR (2004) Public perceptions of unusually warm weather in the UK: impacts, responses and adaptations. *Climatic Research*, 26: 43–59.

Rayner S (2009) Weather and climate in everyday life: social science perspectives. In: Jankovic V and Barboza C (eds.) *Weather, Local Knowledge and Everyday Life*. Rio de Janeiro: MAST: 19–36.

Retrowow (undated). Available from www.retrowow.co.uk/retro_britain/70s/summer_of_76.html (accessed 20 June 2016).

Said EW (2000) Invention, memory, and place. *Critical Inquiry*, 16: 175–192.

Sandbrook D (2012) *Seasons in the Sun: The Battle for Britain, 1974–1979*. London: Allen Lane.

Speller E (2006) The Summer of 76. *Vogue*, July: 151–152.

Stephenson DB (2008) Definition, diagnosis, and origin of extreme weather and climate events. In Diaz HF and Murnane RJ (eds.) *Climate Extremes and Society*. Cambridge: Cambridge University Press: 11–23.

Tebbutt M (2012) Imagined families and vanished communities: memories of a working-class life in Northampton. *History Workshop Journal*, 73 (1): 144–169.

Watkiss Singleton R (2011) *Old Habits Persist: Change and Continuity in Black Country Communities, Pensnett, Sedgley and Tipton, 1945–c.1970*. PhD thesis. University of Wolverhampton.

Wells GL and Loftus EF (2003) Eyewitness memory for people and events. In Goldstein AM (ed.) *Handbook of Psychology: Forensic Psychology*, vol. 11, New York: John Wiley: 149–160.

Williams K (1994) *The Kenneth Williams Diaries*. London: Harper Collins.

3 On the home front

Australians and the 1914 drought

Ruth A. Morgan

Introduction

Just over a decade after the devastating Federation Drought (1895–1903) (Lindesay, 2005), dry conditions again visited the young nation of Australia. By late 1914, drought had stalked through the agricultural areas of the southern half of the continent where it would last long into the following year. In October, just months after the outbreak of the Great War, the Victorian Minister for Public Works Frederick Hagelthorn declared, "The settlers in the mallee are suffering severely to-day, and unless relief comes quickly their position will be quite as bad as, if not worse than, that of the men at the front".[1] By November that year, Western Australia, South Australia, Victoria and southern New South Wales had all been in "the grip of drought" for at least six months.[2] These dry conditions in 1914 were the most intense of a series of dry seasons reaching back to 1911 in South Australia and lingering into 1916 in parts of Victoria (Foley, 1957).

Although a relatively short drought, at least by Australian standards, its wide reach across the southern states proved to be devastating. Just when Australian farmers were supplying the British Empire's war effort, the severity of the drought caused the collapse of the national wheat crop. Averaging less than a bushel and a half per acre, the crop was the worst on record (Dingle, 1984). The coincidence of the drought with the commencement of World War I saw many young rural men enter the armed services in order to escape economic hardship (Welborn, 1982). In Western Australia, these conditions caused the state's wheat yield to plummet by 80 percent in a single year (Powell, 1998). Parishioners in New South Wales prayed for rains, and in South Australia irrigationists lamented the dwindling supplies of the River Murray, which had "ceased to be a river", according to the Chairman of the Victorian Water Commission, Elwood Mead.[3] The extent and severity of these conditions challenged the aspirations of Australians and their governments for the nation's population, progress and prosperity.

According to the Australian Bureau of Meteorology, the 1914 drought was the product of one of the "twelve strongest 'classic' or canonical El Niño events" to affect Australia in the twentieth century (BoM, 2016). The "messy fingerprints" of this global mechanism not only affected parts of Australia, but also caused severe drought in New Zealand, New Mecklenburg (today, New Ireland in Papua New Guinea), South Africa and the Dutch East Indies (Indonesia). Furthermore, this was one of the few occasions where both the south eastern and south western areas of the Australian continent were drought affected (Lindesay, 2005). Originating in the warmer than normal waters of the tropical Pacific Ocean, the El Niño-Southern Oscillation is the primary cause of climatic variability for the eastern half of Australia where it tends to produce particularly dry conditions in El Niño years (BoM, nd). Although the implications of this phenomenon for southern Western Australia are less direct, it tends to result in drier conditions in what is normally a wet winter (Hope et al., 2010).

Recent historical research into Australian climate histories encourages a closer examination of the effects of, and responses to, the 1914 drought in the young nation's wheat regions. This chapter focuses on Western Australia, where the 1914 drought contributed to one of the driest years on the state's record and lingers as a meteorological and cultural marker of the severely dry conditions faced in the state's agricultural areas. From as early as 1915, the drought was framed as both an aberration and opportunity, as a defining experience of character and belonging and as a proxy for predicting the weather. Important to this framing process was the contemporary reportage of local newspapers, which provide insight into how the 1914 drought was perceived and subsequently portrayed. To bolster their assessments, these reports frequently deployed the meteorological records kept by individuals and the state. Close listening to oral history interviews with wheatbelt farmers and their families reveals the extent to which these reports aligned with personal experiences of drought and climate variability in the region. Drawing on these oral histories, meteorological records and newspaper accounts, this chapter examines the ways in which Indigenous and non-Indigenous Western Australians experienced and have remembered this drought as well as how these memories shaped personal and state responses to subsequent periods of water scarcity.

Over the past two decades, environmental historians and historians of rural Australia have turned to memory studies as a means to understand the experience and impact of social and environmental change in regional areas. As Ian Waites shows in his chapter on the British summer of 1976, "high profile weather events" have an especially enduring impact on popular memory. Oral history and the analysis of memory provide insights into personal, family and community narratives, which historians complement with documentary sources, such as newspapers, scientific research, and archival materials (Goodall, 1999). Much of this research was conducted in the context of a sense of 'rural crisis' and decline that dominated regional policymaking in the 1980s and 1990s (Darian-Smith, 2002; McCann, 2005). In the wake of devastating droughts and floods in eastern Australia in the 2000s, combined with growing concerns about anthropogenic climate change, environmental historians are using oral history to elicit the ways Australians respond to and make sense of such dramatic events. Such research seeks to reveal the deeply cultural and personal dimensions of 'natural disasters', as historian Emily O'Gorman observes, "Our understandings of events as disasters are directly influenced by where and how we live and how we understand our places" (O'Gorman, 2012: 3).

In terms of drought and rural Australia, journalist and oral historian Deb Anderson (2014) offers a valuable model for interrogating the ways in which climate is culturally mediated. Anderson's study focused on the semi-arid Mallee region of northwestern Victoria, where she conducted interviews with over 20 residents between 2004 and 2007. These interviews revealed narratives of 'endurance', whereby drought stories formed both "a cultural legacy and a shield from anxieties about the future" (xiv). In contrast to Anderson, this chapter samples oral histories undertaken by the state and national libraries between 1970s and 2000s. Their references to the 1914 drought are entirely unprompted and, as a result, provide unique insights into the way a particular event has informed personal and community narratives of settler identity, belonging and perseverance in the Western Australian wheatbelt since the early twentieth century.

Recording the reliable rainfall of the Western Australian wheatbelt

The Western Australian wheatbelt lies in the state's southwest and extends from the Geraldton sandplains in the north to the mallee of Esperance in the South (Map 3.1). With the Darling Ranges forming its natural western border, the wheatbelt stretches eastward to the continent's arid interior and spans an area about the size of Nebraska, with a population of fewer than 140,000 people (ABS, 2015). The region's average annual rainfall varies from 600 mm in the West to 300 mm in the East. Although there is a paucity of fresh surface water in the region, its relatively flat landscape has made it ideal for grain farming. Since the turn of the twentieth century, successive waves of agriculturists have advanced into this eastern boundary with little regard for the marginal nature of this country or for the Nyoongar people, whose ancestral lands colonists seized for cropping and grazing.

At the turn of the twentieth century, the state's first Premier, Sir John Forrest, began to engineer the development and settlement of the "last of

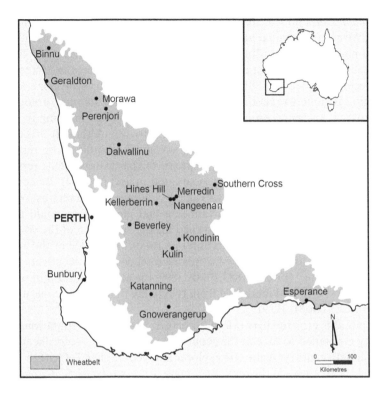

Map 3.1 Map of Western Australia, showing the extent of the wheatbelt.

the world's great wheatbelts" (Murray, 1998: 267). He envisaged agriculture as the stable foundation for economic development and self-sufficiency that the volatile mining industry could not provide (Glynn, 1975). The scale of the endeavour and the difficulties of attracting overseas investment to Western Australia led his government, and those that followed, to assume a central role in the development of the wheatbelt region (Tonts, 2002). Encouraging this agrarian vision in Western Australia was a long-held faith in the reliability of the climate of the state's southwest. The economist William Stanley Jevons had observed in 1859 that this part of Western Australia suffered fewer droughts than the rest of the continent and concluded that it "shows less [sic] variations in the yearly rainfall than the climate of the other colonies" (Jevons, 1859: 60). Jevons had based his judgement on just one year of rainfall records, and 50 years later, rainfall data remained relatively scant.

Although meteorological observations had been kept at the Surveyor General's office between 1830 and 1876, these had not documented that variable so crucial to Western Australia's agricultural progress – rainfall. The commencement of official rainfall records coincided with a series of particularly dry winters in the southwest (Waylen, 1878: 3). The *Western Australian Times* hoped this project would help to "remove the stigma cast upon [Western Australia] at a recent meeting of the Colonial Institute that 'it had never spent a shilling upon investigations of its climate'".[4] But the slow pace of improvement drew the ire of M.A.C. Fraser, the Meteorological Record Keeper. In 1881, he complained to the Surveyor General that, "Nothing is done to record even the bare outline of its meteorology and climatology, a science which at the present time is fast growing in public estimation and importance" (cited in Day, 2007: 10). Although Fraser's report was somewhat exaggerated, meteorological recordkeeping remained limited and restricted to the coastal areas: of the seven stations reporting to his office in 1880, only York was situated inland (1881: 4). By contrast, New South Wales had ten times as many volunteer observers as Western Australia.[5] Attuned to the important role that meteorology could play in agricultural development, Fraser argued for the "adoption of the volunteer system of rain observations ... to obtain a fuller knowledge of the amount of rain that falls annually, how far inland it reaches, and its effects on the yield of crops, etc." (Fraser, 1883: 4). By 1890, the number of rainfall stations reporting to Perth had increased from two to 80, thanks to the recruitment of volunteers (BoM, 1929).

Despite this growing network of observers, the region's reputation for reliability continued to exceed the scope of meteorological recordkeeping in the twentieth century. Antarctic explorer and geographer T. Griffith Taylor extolled the qualities of the region's climate. In *Australia in its physiographic and economic aspects* (1911), Taylor offered his assessment of the economic potential of particular regions of Australia and their viability for white settlement. In Western Australia, he declared, "close settlement is confined to the south-west corner of the continent, which is the only district where good agricultural land suitable for white labour occurs" (108). Two years later, he enthusiastically labelled the region "Westralia felix" in *Climate and Weather of Australia* (1913), published by the recently established Commonwealth Bureau of Meteorology.

"We've only had one drought and that was the 1914": enduring faith in the Western Australian wheatbelt

The onset of drought and war in 1914 was especially challenging to Western Australian farmers and led to grave doubts about the region's suitability for agricultural development, particularly regarding wheat farming in the eastern wheatbelt (Glynn, 1975). These were environmental anxieties that questioned the wisdom of closer settlement and in the case of Western Australia the economic future of the state. As most of the wheatlands had only been settled after 1908, many farmers had struggled to establish themselves under difficult climatic and financial conditions. Compared to the longer established farmers, the newer farmers lacked the capital and experience to cope with the drought. Concerned for the wheatbelt's development, the state government in 1916 established a Royal Commission to inquire into the state's agricultural industries; its findings for the wheatbelt and the southwest coastal areas were published the following year.

Although some witnesses questioned the rainfall records, the Commissioners kept their faith in the southwest's climate. They concluded that, "Comparatively the Grain Belt can be most aptly defined as a second class country with a first class rainfall. Our rainfall ... is more regular than in any of the other States" (1918: xi). According to this logic, the 1914 drought was an aberration in the climate record, a temporary departure from the normal, reliable climate conditions of the region. As an aberration, therefore, the 1914 drought did not dispel the region's rainfall reputation, which was vital to the fulfilment of the state's vision of the wheatbelt's development. Key to this vision was peopling Western Australia, which had long struggled to attract settlers (Morgan, 2015a). According to the Commissioners, "The great need of the State is a large producing population. Every effort should be made, after the war, to encourage to our shores either agricultural labour or those with experience in agricultural pursuits" (1918: xvi).

Framing the 1914 drought as abnormal allowed the event to be recast from a disaster to a character-defining episode for individuals, their families and the state. The decisiveness of this shift was reinforced by the coincidence of the drought with the Great War, which encouraged a militaristic discourse on the home front. For Australian military historian Michael McKernan in his *Drought: the Red Marauder*, "War and drought ... tell us similar things about Australians, "about the spirit and the determination of these remarkable people" (2005: 14–15). In Western Australia, where drought in the agricultural areas was widely believed to be an abnormal phenomenon, failure on the land was not seen to be due so much to the land or climate as to the lack of effort and determination of the farmer (Powell, 1988). The "land is usually good to those who use it well," observed the Royal Commission on the Agricultural Industries in 1917, "while it rejects infallibly the unfit and the ineffective" (xi). Persisting in spite of the drought was, therefore, a sign of virility, resilience and dedication to the state's development – associations that persisted long after the Armistice.

For the state, persistence manifested itself in the improvement of the agricultural technocracy to ensure the wheatbelt's future. Such improvements had been a key recommendation of the Commissioners, who concluded, "It is to better methods of cultivation and the careful selection of seed for different districts that we must look for improvement in our position as a grain-growing State, and it is for this reason that we advocate strongly an efficient Department of Agriculture" (xi). Until the end of the Great War, the eastern wheatbelt was "critically marginal" because of the rudimentary nature of the farming techniques employed in Western Australia at that time (Dahlke, 1975: 4). To assist farmers in their future endeavours, the Royal Commission demanded better services from the state's Department of Agriculture (est. 1894). As the Commissioners argued,

> These men and women pioneers are doing a great national work, often at considerable cost to themselves and their families, and it is the State's duty, and in its own best interests to do everything possible to assist these people in overcoming the natural disadvantages which hamper them and prevent them becoming successful mixed farmers, and as such an asset to the State, instead of cereal growers doomed to a life of hard continuous labour for an inadequate reward.
>
> (xviii)

This was a rural vision that would "preserve the *values* of the pre-industrial world but retain the material benefits of technology" (Murphy, 2009: 125). Investing in the state's agricultural services would, the Commissioners hoped, overcome the prevailing environmental and human limits on the region's economic development.

After the 1914 drought, the national wheat crop bounced back from less than 700,000 tonnes to a record 4.8 million tonnes in 1915 (BoM, 2004). The harvest helped to buoy hopes for ambitious land settlement schemes in the Western Australian wheatbelt, which the state government hoped would relieve post-war unemployment and boost economic development (Glynn, 1975). The combination of the region's recovery and the improvement in the agricultural services of the state provided the foundation for more optimistic responses to later instances of drought, which made explicit comparison to the conditions of 1914.

Heeding the drought of 1914

Western Australian farmers had enjoyed a prosperous decade during the 1920s and many took on debts to expand their farms (Glynn, 1975). On the eve of the new decade, however, commodity prices collapsed with disastrous consequences for the state's wheat farmers. The severity of these economic conditions had left many farmers financially exposed to the onset of dry conditions in the mid-1930s, which lasted until the end of the decade. Aside from 1939, rains across the state's agricultural areas were below average, and the drought of 1940 rivalled that of 1914 in its severity. Invasions of ravenous rabbits, grasshoppers and emus decimated the surviving crops, exacerbating the farmers' plight (Burvill, 1979). For the eastern wheatbelt, circumstances were especially dire – farmers there were over-laden with debt and faced poor seasons, poor soils and poor prices (Maddock, 2009).

Drought had again coincided with world war, and it was with the experience of 1914 that newspapers and politicians made explicit comparison.

The agrarian *Western Mail* observed the similar conditions, "Nineteen hundred and fourteen, the first year of the Great War, remains to this day the yardstick by which southern West Australians judge bad seasons. By a strange coincidence 1940, the first winter of this second Great War, has been the nearest approach within recorded meteorological history to the disaster of 1914".[6] The tenor of these comparative reports was mostly dark. The editorial of the *West Australian* noted, "Since the beginning of this century 1914 has been the only season comparable for rainlessness with 1940 up to the end of August. It seems improbably that this year can prove so disastrous as that memorable earlier drought, but whatever happens it will be bad enough".[7] The *Western Mail* reported, "The 1914 drought was small compared with this one, for it operated over an agricultural area which was not so widely extended as now, and on farms which in many cases were only in the pioneering stage".[8]

Other correspondents were more optimistic, their purpose to boost the morale of those fighting the elements on the home front. In August 1940, for instance, "AHA" mused in the *Gnowerangerup Star*, "I should be able to give some measure of hope and encouragement so such people on the principle that we can very often judge the future by the past, and what has happened once may happen again, by telling my experiences during the driest season I have known in the thirty-six years just past".[9] Rallying their readers, newspapers reminded them that 1914 had shown that drought was not a cause for gloom. As AHA continued, "So cheer up you down hearted ones, you may get a good rain yet, sufficient to mature your crops and all your dams". In the *Western Mail*, "Wheat Ears" of Bunbury was similarly hopeful, "It's not always easy to be cheerful when up against trouble but it's a help to look forward to better things ahead. And our experience was that there were better times ahead".[10]

Correspondents echoed the findings of the Royal Commission that followed the 1914 drought. They were at pains to highlight the infallibility of the land and the region's reliable weather. The *West Australian* reported, "Seasons like this are fortunately rare in Western Australia",[11] while Wheat Ears lamented,

> I often think we hear so much of the distress and want in the wheatbelt and no mention is made as a rule of those who have done well and have reaped many benefits. I think in many cases of failure to make good, it is not the fault of the land but the fact that many people holding land do not understand the handling of it.[12]

The skills and mentality of local farmers were again key to the way they experienced and responded to drought. The *Western Mail* urged the rural readership to see the opportunities for improvement that the drought conditions offered, "The 1914 drought taught Western Australia the value of fallow. The 1940 drought is teaching farmers the value of fodder reserves.

It has also emphasised the urgency of enlarging water storages and improving country water catchments".[13] In addition to these morality tales,
local newspapers drew on meteorological records to clarify their comparison of the 1914 and 1940 droughts. C.J. Bull of Kulin, who had kept rainfall records for the Perth Observatory since 1913, shared his data with the
West Australian. According to his records, "bad as this season has been in
this district, 1914 was far worse".[14] Similarly, T. Hicks' rainfall record at
Beverley showed that the rainfall for 1940 had been "considerably higher
than for the drought year of 1914".[15] Although newspapers conceded that
elsewhere the 1940 drought had been more severe than 1914, these examples provided quantitative evidence that drought had spared some areas,
and therefore the wheatbelt's reputation for reliable rainfall was not entirely tarnished. As the *Western Mail* asserted over Christmas in 1940,
"Taking the long view, Western Australia may claim to be singularly free
from severe droughts in the agricultural, dairying and fruitgrowing areas
of the state".[16]

Drought and belonging in the Western Australian wheatbelt

In subsequent decades, personal meteorological records continued to provide a local perspective on prevailing weather conditions. From 1940, the
weekly *Beverley Times* newspaper (1905–1977) regularly cited the rainfall
records of Ossie Vallentine (b. 1908) of Redlands, west of Beverley, where
his father had commenced recordkeeping in 1910. Redlands became an
official station of the Bureau of Meteorology the following year (BoM,
2015). As the records had been "meticulously kept for a very long period", Vallentine's data was regularly summoned in the 1950s and 1960s
to contextualise particularly dry conditions for readers in the newspaper's
distribution area in the western central wheatbelt, which extended from
Talbot Brook in the west to Badgin in the north, Quairading in the east,
and Brookton in the south.[17] Historic and current weather data were analysed together, as the *Beverley Times* noted, "The year 1914 figures largely
in the agricultural history of W.A. as the only real drought year for these
areas – and all years of low registration appear to be inevitably compared
with the period".[18]

The regular publication of Vallentine's rainfall record performed at least
two functions. For the Shire of Beverley, the data substantiated faith in the
reliability of the local climate. In the drought year of 1969, the *Beverley
Times* proudly observed, "The Dale portion of the Beverley district is once
again the most favoured area in the present season of drought conditions
over much of the State and this is evident from a perusal of figures kindly
submitted by Mr Ossie Vallentine".[19] Elsewhere, farming shires were reluctant to proclaim their area drought-stricken. As the Minister for Agriculture reported to Parliament in early August 1969, "Not one local authority
will admit that drought conditions exist".[20] Although such a proclamation

entitled farmers to drought assistance from both the Commonwealth and state governments, these shires were concerned that the stigma associated with the label of "drought-affected" might damage their shire's reputation as a profitable farming area and cause land prices to plummet.[21] The reality of drought clearly conflicted with long-held official and lay ideas about the wheatbelt's climate.

In addition to upholding the climatic reputation of the local shire, sharing his rainfall record was of significant benefit to Ossie Vallentine. In an oral history interview conducted by the Battye Library of West Australian History in 1995, Vallentine's wife, Sylvia, described her late husband as a hardworking, successful farmer and keen sportsman of aristocratic heritage. Vallentine was also a "great raconteur" and "president of all these various [local] clubs", which suggests he was a prominent person in the Beverley community (Vallentine, 1995). Sharing his rainfall records with the *Beverley Times* was another means to maintain his status in the region as the descendant of an important local family and as a meteorological authority with state recognition. For Vallentine, therefore, his lengthy rainfall records served as a quantitative chronicle of his family's history in Beverley and their contribution to the area (see Davison, 2000).

The onset of drought conditions in 1969, which severely affected farmers across the wheatbelt, foreshadowed the dry seasons that characterised the following decade. Although many farmers found the seasonal conditions of the early 1970s difficult, the series of dry years later in the decade were unprecedented in their severity and impact. Unusually low winter rains in 1976 brought drought conditions to many areas of the state's southwest, which lasted into the early 1980s for many farmers, particularly in the north eastern wheatbelt (Department of Agriculture, 1983). The severity of the drought shocked farmers across the region, with old-timers like Dalwallinu farmer Ted Black noting that he had "never seen it as dry".[22] As these conditions lingered, farmers in the north eastern wheatbelt suffered four consecutive years of drought. Even areas like Kondinin and Katanning, where rainfall had been considered safe and reliable, were affected.[23] As Perenjori farmer Bill Bestry recalled in the late 1990s, "They were our hardest years ... They set us back a lot, set us back a long way, those years" (Bestry, 1998). By the end of the decade, the drought-affected area extended from the north eastern wheatbelt and upper central region down into the southern wheatbelt.[24]

The droughts of the 1970s coincided with a renewed emphasis on local history, which coalesced around the sesquicentenary of Western Australia's colonial foundation in 1829. Such coincidences help to embed extreme events in collective memory and regional history, as Alexander Hall shows in his chapter on flooding in northern England. Among the initiatives commemorating the state's 175th anniversary was a concerted effort by the Library Board of Western Australia to undertake a program of oral history interviews with local 'ordinary' working people.

These interviews, which undertook a life history approach, offer a personal lens through which to view the state's twentieth-century history. In some of the interviews undertaken in 1978, which focused on rural life in the Western Australian wheatbelt, the 1914 drought is deployed as an example of the hardship families overcame. David Giles (b. 1899) of Hines Hill, whose family moved there from New South Wales in 1909, recalled, "[1911] was one of the driest years ... 1912 was a mediocre year and 1913 wasn't very much better and '14 was a drought. They had to import stock feed from India, imported corn from India, terrible". Subjects emphasised their family's self-sufficiency and independence during this difficult period. For Albert Rutter (b. 1905), "[T]he 1914 they got nothing, there was complete [crop] failure ... In the early days there was no banks, no sustenance, or no help whatsoever. They had to look after themselves and grow their own things". Describing his father's efforts, Vincent Cahill (b. 1903) observed, "it kept us going ... we were never on the banks". As the Cahill and Rutter families had both established themselves in Nangeenan in 1904, the 1914 drought and the preceding dry seasons challenged these farming families in the first decade of their lives in the Western Australian wheatbelt.

The biographical nature of these interviews facilitated the subjects' particular framing of their own life histories. For these subjects, the 1914 drought was a significant life event around which they could curate a suite of personal characteristics that shaped a particular identity. The identity that these subjects conjure is akin to the 'pioneer'. In the Australian context, according to historian John Hirst, the pioneer is a character defined by "courage, enterprise, hard work and perseverance; it usually applies to the people who first settled the land" (1978: 316). The so-called 'pioneer legend', which had prevailed in Australian popular culture since the late nineteenth century, was a nationalist myth that celebrated individual endeavour. The Western Australian Premier Sir Charles Court, for instance, declared at a sesquicentenary event, "Western Australians of all generations had done a mighty good job in a short time ... We have done it by work and enterprise and by backing ourselves to win against great odds" (cited in Bolton, 1989: 16). Although these interview subjects were but youths during the 1914 drought, their families' struggles were evidence that they themselves were pioneers of the wheatbelt.

Ironically, the reminiscences of these Western Australians coincided with an emerging critique of this 'pioneer legend' as a conservative mythology that sustained a white, patriarchal, classless view of Australian history (see Hirst, 1978; Lake, 1981). In his 1985 critique of the pioneer legend in the Western Australian context, social historian Tom Stannage argued that this narrative had produced a "gross distortion of the reality of the past" that overlooked the state's history of Aboriginal dispossession and racial inequality (150). The infusion of the subjects' accounts with this pioneer mythology points to what historian Bain Attwood suggests is a performative means to legitimate settler possession. Telling stories not only serves to "represent possession but also to re-enact it" (2009: 105). In his interview, for example, Giles remembered,

> When we first came here the aboriginals were camped here where the recreation ground is now. ... I don't know how many was here, but it was quite a lot. Some worked on the farms, but they'd go and catch 'roos and things like that mostly, they didn't do work round the place. ... We treat them like dogs and they have to act like dogs.

This derogatory portrayal of Aboriginal people in Hines Hill contrasts starkly with Giles' appraisal of the white farming community. According to his account, the traditional owners are unworthy of the wheatbelt lands, while farmers have a rightful claim to the land as a consequence of their labour and endeavour. This mindset echoed another Australian mythology, that cultivation was equated with ownership and possession. For many of the Nyoongar (the Aboriginal people of the southwest), the 1914 drought had played a significant role in their dispossession from the land. A year prior to the onset of the 1914 drought, the Commissioner of the Wheatbelt, George L. Sutton, presented a lecture to an audience at the Western Australian Museum. Using lantern slides, he showed his audience the state's agricultural progress and explained, "Yesterday our wheat lands were a blacks' camp. To-day they are being broken up with the most suitable implements modern engineering can devise".[25]

Sutton's speech deployed the rhetoric of white development: that agricultural cultivation was a moral act of civilisation that rendered white (Western) Australians superior to Aboriginal peoples, who seemed to lack the ability to till the soil. Yet many Aboriginal people remained in the emerging agricultural areas of Western Australia. Historian Anna Haebich estimates that about three-quarters of the southwest's Aboriginal population (possibly 1,500 people) lived in the wheatbelt region at the turn of the twentieth century (1984: 61).

Since the dry conditions of 1911, many Aborigines had moved into camps on the outskirts of wheatbelt towns (Haebich, 1992). In one town, the Aboriginal population increased fivefold in three years, from 40 to over 200 (Haebich, 2004). Accompanying this shift was a threefold increase in the reliance of Aborigines on government rations, from about 1,000 in 1907 to over 3,000 by the outbreak of the Great War (Haebich, 1992). As Haebich explains, the state of the camps quickly degenerated: "There were no proper shelters, no sanitary or rubbish services, no fresh water, no work and only meagre rations of flour, tea and sugar for the elderly and dependent mothers, issued by the police on behalf of the Aborigines Department" (2004: 272–274). Many perished as a result of these conditions (Haebich, 2004: 272–274). The development of the Western Australian wheatbelt and the onset of drought, therefore, forced many Aborigines onto the fringes of the white settlements of the wheatbelt. Although there are few (if any) extant Aboriginal recollections of the 1914 drought, Haebich offers some insight from her own research. Writing five years after the 1979 sesquicentenary celebrations and the state library's oral history interviews, she noted that, "The descendants of some Aboriginal farmers believe they were 'tricked' out of the land by local white farmers or, unable to keep up with the necessary payments, they simply walked off the land" (1984: 66).

"Worse than 1914": a changing climate

At the dawn of the new millennium, drought crept across the wheatbelt's paddocks once again. As the dry conditions wore on, government officials declared that the agricultural areas faced the region's worst drought on record.[26] The onset of the drought hit some farmers especially hard. Many were still recovering from devastating episodes of frost and locust plagues in the late 1990s. For the farming sector, comparisons with 1914 strengthened their claims for Commonwealth drought assistance. The state agriculture minister observed in July 2001, "Rainfall records indicate that 1914 was the worst drought this State has ever had. But this is worse than 1914".[27] Charles Hyde, a third-generation Dalwallinu farmer, whose family had been on the property in 1914, felt that "We have never had it this bad".[28] The severity of the drought qualified many affected farmers for relief under the national drought policy provisions for 'exceptional circumstances'.[29]

In the capital city of Perth, these dry conditions manifested in what the state government described as a "water crisis". Comparisons to the 1914 drought highlighted the severity of the situation. In August, one government minister told Parliament, "The estimated inflow this year [into local dams] has been just 13 million kl compared with 21 million kl that flowed in during the drought of 1914, previously the lowest on record".[30] In the following months, local newspapers repeated that conditions represented "one of the driest periods on record", encouraging households to reduce their water consumption.[31] In April the following year, the director of the state's water utility warned, "We are in a very serious situation, with the worst winter since 1914 and Perth dams at the lowest levels ever".[32]

Coinciding with the onset of the drought, Australian climate researchers had concluded that Perth, the wheatbelt and the surrounding southwest region had experienced an ongoing drying trend since at least the mid-1970s, which they attributed to anthropogenic climate change (IOCI, 2002). In 2004, the mammologist Tim Flannery predicted that, "Perth will become a ghost metropolis over the next few decades unless governments acknowledge that global warming is a reality".[33] The prevailing dry conditions (and associated restrictions on household water use) helped substantiate this prediction, which fed concern about the prospect of very scarce water resources in the future. These urban anxieties about ongoing water scarcity encouraged a program of aggressive water resource development for the city, including the construction of the nation's first seawater desalination plant in 2006 (Morgan, 2015b).

Although the comparison with 1914 had contributed to urban concerns about anthropogenic climate change, rural Western Australians were more ambivalent. At the height of the Millennium Drought (1997–2010), the National Library of Australia dispatched a team of researchers in 2008 to interview rural Australians living in drought-affected areas around the country. In Western Australia, local historian and broadcaster Bill Bunbury spoke with seven farming families about their experience of the ongoing drought and their prospects for the future. Although the 1914 drought was beyond the living memory of the subjects, the event was discussed in four of these interviews to describe the relative severity of the prevailing drought and say whether it was part of a broader pattern of climate change.

In Morawa, the Collins family saw a trend in the local rainfall record. Gary (b. 1954) reflected, "In past times we did have dry seasons ... but they didn't last for any length of time. Like the 1914 drought, there was 1922. And various situations like that. One or two year events, very quickly recovered. ... Records are telling us there we're drying. ... This seems to be a trend we may have to accept". For the Collins family, the run of dry seasons suggested a changing climate. According to a 2006 study conducted by the Australian Bureau of Statistics (ABS), their views aligned closely with others in the northern wheatbelt, where over 80 percent of farmers thought that the local climate had changed and was affecting their holding, and 70 percent of farmers had changed their management practices as a result. Elsewhere, however, about 50 percent of farmers thought the local climate had changed, and just a third had changed their farming practices. Whether these changes in climate were anthropogenic or natural was not in question (ABS, 2008).

To the other subjects, the 1914 drought suggested a different view. Despite his wife Helen's protests, John Nankivell (b. 1935) of Wubin defended the local rainfall record by declaring, "We've only had one drought and that was the 1914". For Murray Criddle (b. 1943) of Binnu and Bob Panizza (b. 1936) of the Yilgarn, the 1914 drought suggested the current conditions were within the normal range of experience. Recalling his father's experience of the 1914 drought while farming at Dindaloa, Criddle suggested, "there were always dry periods". Panizza observed that the current drought was "part of a larger pattern": "[W]e've had two lean years, but we've had two lean years back in 1914/15. We've seen them back in 1947/48, we've seen it '56/57 ... very seldom ever goes into a third year". These subjects were reluctant to interpret the prevailing dry conditions as part of a broader trend of climate change, natural or otherwise. As Panizza observed, "We'd need to get a different pattern to say right, climate change is definitely with us".

That the climate conditions elicited such mixed responses points to not only the unevenness of the drought's impacts, but also to the political divisiveness of the climate change issue in Australia when the interviews were undertaken. A 2008 survey of over 250 Western Australian farmers found that only a third of survey participants agreed that climate change was occurring, and just 19 percent believed climate change was human induced. In their discussion of these results, the researchers noted that many participants found "the question was confronting, or expressed a reticence to respond". When given the option to attribute climate change to natural causes, they "exhibited relief". Most doubted that climate change policy would be sensitive to the needs of their communities and expected politicians to "climate change as an election issue" (Evans et al., 2011: 217–235). Already the 2007 Federal election had been tightly contested with climate change a key concern of the campaign. The success of the Australian Labor Party brought a greater political will to address the matter after the decade-long leadership of the conservative Coalition, which had refused to ratify the Kyoto Protocol (Rootes, 2008).

In the shadows of the climate change debate hid another issue: the future of Australia's rural sector. The sector had undergone significant transformation in the 1980s and 1990s, and challenges farmers faced as a result were not only economic but also cultural and political, as Australians collectively lost their sense of "countrymindedness" (see Aitkin, 1985; 2008). The forces of globalisation were taking their toll on farming communities as pressures mounted for farm amalgamation, greater efficiency and productivity gains, and increased involvement with agribusiness (Lawrence, 2005). These changes compounded the persistent problems of technological change, environmental degradation and rising debt, accelerating a drift to the cities, economic hardship and a decline in morale. Historian Graeme Davison notes that while none of these challenges was unique, "What was new was the strength of the combined force with which they now acted, and the changed framework of expectations in which their impact was now interpreted" (2005: 53). In terms of the national drought policy, as historian Judith Brett has observed, "Drought-stricken farmers were no longer heroic victims of fickle nature, but merely bad risk managers" (2011: 49). Climate change, natural or otherwise, was yet another challenge facing farmers in the Western Australian wheatbelt.

Associated with the changing landscape of Australian farming was a rural scepticism of government authority and science, particularly with regard to climate. The Bureau of Meteorology's inability to provide long-term or seasonal forecasts to wheatbelt farmers had long been a source of discontent. As a result, most farmers remained dependent on their own local knowledge and the Bureau's short-term forecasts to manage the seasons (Morgan, 2015b). In the oral history interviews, references to the rainfall records of their properties, the multigenerational nature of these records, and the weather patterns observed over time all highlighted the importance of local weather knowledge for these farmers (Anderson, 2014). The researchers associated with the 2008 survey of Western Australian farmers concluded that the longer farmers had been involved in farming, the more likely they were to dismiss the scientific research on climate change (Evans et al., 2011). Invoking the 1914 drought was a means then to emphasise their authority on matters of local climate change and their commitment to the land, and to recall what they believed to be a golden era of farming in Western Australia.

Conclusion

In 2010 the Millennium Drought broke, bringing relief to farmers across southern and eastern Australia. In the Western Australian wheatbelt, however, 2010 was the driest year since records commenced, with some areas observing half, or less, of the average annual rainfall (BoM, 2011). Forecasts suggest that such conditions will only continue as a result of the ongoing regional drying trend and rising temperatures. The media has largely portrayed their plight in Steinbeckian terms, depicting scenes of rural decline and exodus. National reforms to drought assistance might come too late for some of them. But many farmers are fighting on: a ten-year study of broadacre (large-scale grain and/or livestock) farming in Western Australia found that nearly two-thirds of farm businesses are in a growing or strong position (Kingwell et al., 2013). These farms, the study reported, are largely dependent on wheat growing, and their profitability is the product of improving productivity, training and technical efficiency – the cornerstones of Australian agricultural policy since at least the 1990s.

Despite the dramatic changes in Western Australian farming since the turn of the twentieth century, the continued potency of the 1914 drought as a meteorological and cultural marker of rainfall suggests that the personal and community narratives of identity and belonging remain more important than ever. For the state, comparisons with the 1914 drought have been a means to articulate the severity and hardship of the present, while serving as a wellspring of perseverance. The use of oral histories in combination with documentary sources provides insights into the intimate ways that past weather events significantly shape understandings and experiences of prevailing conditions. For the participants in the oral history interviews undertaken in the 1970s, 1990s and 2000s, the 1914 drought was an experience of their parents or grandparents on which they could draw for strength at times of climatic and economic uncertainty. The continued resonance of extreme weather events across generations underscores the importance of recognising the ways in which climate change and variability manifest culturally in the past, present and the future.

Acknowledgements

This chapter has benefitted enormously from the generous participant feedback of the conferences "Disasters Wet and Dry", Renmin University of China, 23–26 May 2013 (especially Nick Breyfogle for his insightful commentary) and "The Australian natural environment as a threatening and threatened entity", University of Tübingen, 15–16 August 2013 (especially Sabine Sauter, Cameron Muir and Luke Keogh). The author is also grateful for the advice of her research group at Monash University, particularly Bain Attwood and Kate Murphy, and for the guidance of Andrea Gaynor, University of Western Australia. Map 3.1 produced courtesy of Kara Rasmanis, Monash University.

Notes

1 *Kerang New Times* 2/10/1914, 6.
2 *Leader* 28/11/1914, 6.
3 *Argus* 5/12/1914, 20.
4 *Western Australian Times* 22/3/1878, 2.
5 *West Australian* 1/2/1886, 3.
6 *Western Mail*, 26/12/1940, 89.
7 *The West Australian*, 6/9/1940, 12.
8 *Western Mail*, 26/9/1940, 27.
9 *Gnowerangerup Star*, 10/8/1940, 2.
10 *Western Mail*, 12/12/1940, 15.
11 *West Australian*, 6/9/1940, 23.
12 *Western Mail* 12/12/1940, 15.
13 *Western Mail*, 26/12/1940, 89.
14 *West Australian*, 14/11/1940, 14.
15 *West Australian* 15/1/1941, 3.
16 *Western Mail*, 26/12/1940, 89.
17 *Beverley Times* 3/6/1966, 1.
18 *Beverley Times*, 8/8/1969, 1.
19 *Beverley Times*, 5/9/1969, 1.
20 *WAPD* 6/8/1969, 88.
21 *Countryman* 21/8/1969, 9.
22 *Western Farmer* 15/7/1976, 1.
23 *Western Farmer* 18/5/1976.
24 Drought Consultative Committee 14/5/1981.
25 *West Australian* 1913, 7–8.
26 *West Australian* 18/7/2001, 7.
27 *West Australian* 7/7/2001, 8.
28 *Sunday Times* 29/9/2002, 3.

29 *West Australian* 12/11/2002, np.
30 *West Australian* 1/8/2001, np.
31 *Sunday Times* 18/11/2001, 2.
32 *West Australian* 4/4/2002, np.
33 *West Australian* 25/6/2004, np.

References

Aitkin D (1985) 'Countrymindedness' – the spread of an idea. *Australian Cultural History*, 4: 34–41.

Aitkin D (2008) Return to 'countrymindedness'. In Davison G and Brodie M (eds.) *Struggle Country*. Melbourne: Monash University Press: 11.1–11.6.

Anderson D (2014) *Endurance*. Collingwood, VIC: CSIRO Publishing.

Argus, 5 December 1914, p. 20.

Attwood B (2009) *Possession: Batman's Treaty and the Matter of History*. Melbourne, VIC: Miegunyah Press.

Australian Bureau of Statistics (ABS) (2008) *Farm Management and Climate, 2006–2007*, no. 4625.0. Canberra, ACT: ABS.

Australian Bureau of Statistics (ABS) (2015) *National Regional Profile: Western Australia – Wheat Belt (SA4)*. Canberra, ACT: ABS.

Bestry WP (1998) Interviewed by J. Bannister, 30 March 1998, OH2885, SLWA.

Bolton G (1989) WAY 1979: whose celebration? *Studies in Western Australian History*, 10: 14–20.

BoM (nd) BoM, Climate Glossary, nd, www.bom.gov.au/climate/glossary/elnino.shtml (accessed 10 June 2016).

Brett J (2011) Fair share: country and city in Australia. *Quarterly Essay*, 42: 1–67.

Bureau of Meteorology (BoM) (1929) *Results of Rainfall Observations Made in Western Australia*. Melbourne, VIC: H. J. Green.

Bureau of Meteorology (BoM) (2004) *Drought, Dust and Deluge: A Century of Climate Extremes in Australia*. Canberra, ACT: Bureau of Meteorology.

Bureau of Meteorology (BoM) (2011) *Western Australia in 2010: a very dry year in southwest Western Australia*, 4 January 2011, www.bom.gov.au/climate/current/annual/wa/archive/2010.summary.shtml (accessed 10 June 2016).

Bureau of Meteorology (BoM) (2015) *Basic climatological station metadata*, 23 November 2015, www.bom.gov.au/clim_data/cdio/metadata/pdf/siteinfo/IDCJMD0040.010634.SiteInfo.pdf.

Bureau of Meteorology (BoM) (2016) *Australian rainfall patterns during El Niño events*, www.bom.gov.au/climate/enso/ninocomp.shtml (accessed 10 June 2016).

Burvill GH (1979) *Agriculture in Western Australia*. Nedlands, WA: UWA Press.

Cahill V (1978) Interviewed by M. Adams, 1978, OH281, SLWA.

Collins G, Collins D and Collins J (2008) Interviewed by B. Bunbury, 2 April 2008, TRC5945/7, National Library of Australia (NLA).

Countryman, 21 August 1969, p. 9.

Criddle M (2008) Interviewed by B. Bunbury, 3 April 2008, TRC5945/9, NLA.

Dahlke J (1975) Evolution of the wheat belt in Western Australia: thoughts on the nature of pioneering along the dry margin. *Australian Geographer*, 13: 3–14.

Darian-Smith K (2002) Up the country: histories and communities. *Australian Historical Studies*, 33 (118): 90–99.

Davison G (2000) *The Use and Abuse of Australian History.* Crows Nest, NSW: Allen & Unwin.

Davison G (2005) Rural sustainability in historical perspective. In Cocklin C and Dibden J (eds.), *Sustainability and Change in Rural Australia.* Sydney, NSW: UNSW Press: 38–55.

Day D (2007) *The Weather Watchers.* Carlton, VIC: Melbourne University Publishing.

Department of Agriculture (WA) (1983) *Annual Report 1983.* Perth, WA: Department of Agriculture.

Dingle T (1984) *The Victorians: Settling.* McMahons Point, NSW: Fairfax, Syme and Weldon Association.

Drought Consultative Committee (1981) "Drought in WA 1980/81", 14 May 1981, Cons 7203, V126, State Records Office of Western Australia.

Evans C, Storer C and Wardell-Johnson A (2011) Rural farming community climate change acceptance: impact of science and government credibility. *International Journal of Sociology of Agriculture and Food,* 18 (3): 217–235.

Foley JC (1957) *Droughts in Australia: Review of Records from Earliest Years of Settlement to 1955.* Melbourne, VIC: Bureau of Meteorology.

Fraser MAC (1881) Meteorological Report for the Year 1880, in *WALCV&P,* Perth: Govt Printer.

Fraser MAC (1883) Meteorological Report for the Year 1882, in *WALCV&P,* Perth: Govt Printer.

Giles G (1978) Interviewed by M. Adams, 1978, OH273, SLWA.

Glynn S (1975) *Government Policy and Agricultural Development: A Study of the Role of Government in the Development of the Western Australian Wheatbelt, 1900–1930.* Nedlands, WA: UWA Press.

Goodall H (1999) Telling country: memory, modernity and narratives in rural Australia. *History Workshop Journal,* 47: 161–190.

Haebich A (1984) European farmers and Aboriginal farmers in south Western Australia. *Studies in Western Australian History,* 8: 59–67.

Haebich A (1992) *For Their Own Good: Aborigines and Government in the Southwest of Western Australia, 1900 to 1940.* Nedlands, WA: UWA Press.

Haebich A (2004) 'Clearing the wheat belt': erasing the Indigenous presence in the southwest of Western Australia. In Dirk Moses A (ed.) *Genocide and Settler Society: Frontier Violence and Stolen Indigenous Children in Australian History.* New York: Berghahn: 267–289.

Hirst J (1978) The pioneer legend. *Historical Studies,* 18, 71: 316–337.

Hope P, Timbal B and Fawcett R (2010) Associations between rainfall variability in the southwest and southeast of Australia and their evolution through time. *International Journal of Climatology,* 30 (9): 1360–1371.

Hunt HA, Taylor GT and Quayle ET (1913) *The Climate and Weather of Australia.* Melbourne, VIC: Government Printer.

Indian Ocean Climate Initiative (IOCI) (2002) *Climate Variability and Change in South West Western Australia.* Perth, WA: IOCI.

Jevons WS (1859) Some data concerning the climate of Australia and New Zealand. In *Waugh's Australian Almanac for the year 1859.* Sydney, NSW: James William Waugh: 60.

Kingwell R, Anderton L, Islam N, Xayavong V, Wardell-Johnson A, Feldman D and Speijers J (2013) *Broadacre Farmers Adapting to a Changing Climate*. Gold Coast, QLD: NCCARF.

Lake M (1981) Building themselves up with aspros: pioneer women reassessed. *Hecate*, 7 (2): 7–19.

Lawrence G (2005) Globalisation, agricultural production systems and rural re-structuring. In Cocklin C and Dibden J (eds.) *Sustainability and Change in Rural Australia*. Sydney, NSW: UNSW Press: 104–120.

Leader (Melbourne, VIC) 28 Nov 1914, p. 6.

Lindesay JA (2005) Climate and drought in the sub-tropics: the Australian example. In Botterill LC and Wilhite DA (eds.) *From Disaster Response to Risk Management*. Dordrecht: Springer: 15–36.

Maddock J (2009) Marginal areas. In Gregory J and Gothard J (eds.) *Historical Encyclopedia of Western Australia*. Crawley, WA: UWA Press: 552–553.

McCann J (2005) History and memory in Australia's wheatlands. In Davison G and Brodie M (eds.) *Struggle Country: The Rural Ideal in Twentieth Century Australia*. Clayton, VIC: Monash University Publishing.

McKernan M (2005) *Drought: The Red Marauder*. Crows Nest, NSW: Allen & Unwin.

Morgan RA (2015a) Salubrity and the Swan River Colony. In Varnava A (ed.) *Imperial Expectations and Realities: El Dorados, Utopias and Dystopia*s. Manchester: Manchester University Press: 89–104.

Morgan RA (2015b) *Running Out? Water in Western Australia*. Crawley, WA: UWA Publishing.

Murphy K (2009) 'The modern idea is to bring the country into the city': Australian urban reformers and the ideal of rurality, 1900–1918. *Rural History*, 20 (1): 119–136.

Murray D (1988) Land settlement and farming systems. In Hunt L (ed.) *Yilgarn: Good Country for Hardy People—The Landscape and People of the Yilgarn Shire, Western Australia*. Southern Cross, WA: Shire of Yilgarn: 267–350.

Nankivell J and Nankivell H (2008) Interviewed by B. Bunbury, 3 April 2008, TRC5945/10, NLA.

O'Gorman E (2012) *Flood Country*. Collingwood, VIC: CSIRO Publishing.

Panizza R Interviewed by B. Bunbury, 8 March 2008, TRC5945/4, NLA.

Powell JM (1988) *An Historical Geography of Modern Australia: The Restive Fringe*. Cambridge: Cambridge University Press.

Powell JM (1998) *Watering the Western Third*. Perth, WA: Water and Rivers Commission.

Rootes C (2008) The first climate change election? The Australian general election of 24 November 2007. *Environmental Politics*, 17 (3): 473–480.

Royal Commission on the Agricultural Industries of Western Australia, *Progress Report of the Royal Commission on the Agricultural Industries of Western Australia on the Wheat Growing Portion of the Southwest Division of the State*, Perth: Govt. Printers Office, 1917.

Royal Commission on the Agricultural Industries of Western Australia, *Second Progress Report of the Royal Commission on the Agricultural Industries of Western Australia on the Wheat Growing Portion of the Southwest Division of the State*, Perth: Government Printers Office, 1918.

Rutter A (1978) Interviewed by M. Adams, 29 January 1978, OH274, SLWA.

Stannage T (1985) Western Australia's heritage: the pioneer myth. *Studies in Western Australian History*, 29: 145–56.

Taylor TG (1911) *Australia in its Physiographic and Economic Aspects*. Oxford: Clarendon Press.

Tonts M (2002) State policy and the yeoman ideal: agricultural development in Western Australia, 1890–1914. *Landscape Research*, 27 (1): 103–115.

Vallentine SC (1995 and 1996) Interviewed by Stuart Reid, Oct 1995 and April 1996, OH2693, SLWA.

Waylen AR (1878) Report by the Colonial Surgeon on the Public Health of the Colony for the Year 1877, in *WALCV&P*, Perth, Government Printer.

Welborn S (1982) *Lords of Death: A People, A Place, A Legend*. Fremantle, WA: Fremantle Arts Centre Press.

4 Extreme weather and the growth of charity

Insights from the Shipwrecked Fishermen and Mariners' Royal Benevolent Society, 1839–1860

Cathryn Pearce

Introduction

There has been a call within historical geography for a vision that "puts seas and oceans at the centre of its concerns" (Lambert et al., 2006: 479) to think about the world in ways other than terrestrial. By bringing 'water worlds' into the centre of our analyses, new forms of knowledge and ways of thinking can bring greater understanding (Anderson and Peters, 2014). This call is concomitant with moves in other disciplines to become more sea- and environment-centred, including in history (Bolster, 2006, 2008; Land, 2007; O'Hara, 2009; Williams, 2010; Rozwadowski, 2013) and in literature (Mathiesen, 2016) to create a 'maritime/coastal turn' in the social sciences and humanities.

Climate history research has benefitted by taking a longitudinal, maritime perspective using the skills and sources of historians and historical geographers. Ships' logs have been valuable to study sea level pressure (Küttel et al., 2010), the extent of sea ice and wider climatic patterns (Wheeler and García-Herrera, 2008; Wheeler, 2014). Ships' logbooks are also being used in the 'Old Weather' citizen science project for the United States' National Oceanic and Atmospheric Administration (NOAA), begun in 2010. These sources have also been key to the decade-long multidisciplinary international study, the History of Marine Animal Populations (HMAP), part the Census of Marine Life Project (2000–2010), which discovered new species as well as important data about the population of past species and their ecosystems (Holm et al., 2001).

These initiatives have brought into focus the importance of considering the role and agency of the climate, seas and oceans on human activity and of understanding that "the sea has its own nature" (Anderson and Peters, 2014: 10). It is a space that is unpredictable, volatile and ever changing; humans have had to adapt constantly to the vagaries of the sea for their very survival. These adaptations were not only played out technologically, but also in social, cultural, and political spheres that have normally been restricted to a 'land-based' perspective. As Anderson and Peters argue (2014: 15), "the sea *touches* our everyday lives" and "alerts us to the material and tangible reality of water worlds".

The reality of these 'water worlds' is, of course, that they can be dangerous and life taking as well as life sustaining. On Sunday, 28 September 1838, for example, 11 fishing boats, with 26 men, left Clovelly harbour for what should have been a routine evening's fishing. The weather had been fine when they set out, but by midnight, it had worsened. A 'hurricane,' as the *North Devon Journal* reported, blew up and "raged with so much violence that it left them no means of escape"; nine boats containing 22 men were lost.[1] The villages of Clovelly, Buck's Mill, Ilfracombe, Appledore, Bideford, Hartland and Bude were affected. The storm left 14 widows, 15 children and six 'families'[2] without fathers. Although damage was widespread, no other area suffered quite the same scale of loss as had north Devon's fishing villages.[3]

Traditionally, such accounts of shipwrecks and fishing disasters have been used to create narratives of pathos and the sublime (Corbin, 1994) or tales of gallantry (Gleeson, 2014), but the records can also be used with a different frame of reference, to understand past sociocultural developments and even wider national social movements, as people responded to the impact of climatic events. At a micro-level, however, shipwreck records offer a way to envisage how individual coastal communities experienced their natural environment, which occasionally consisted of gales that became particularly violent. Moreover, these records can also offer insights into this relationship between humans and the sea and can extend the study of charity and philanthropy into a maritime dimension. The storm of 28 September 1838 noted above and the loss of the Clovelly fishing fleet was the inspiration for the establishment of the Shipwrecked Fishermen and Mariner's Royal Benevolent Society (SMS). It represents an extreme weather event that has thus been engrained within institutional memory.[4]

Extreme weather events also galvanised a wider-scale, state-driven response with the institution of a coastal weather warning system by the British Meteorological Office under Captain Robert FitzRoy after the 'Royal Charter' storm of 1859 (Burton, 1986; Anderson, 2005). However, during the nineteenth century, state involvement was limited by the ideology of laissez-faire and free trade. Societal responses to extreme weather and to other socioeconomic concerns were left to individuals, churches and local societies. Consequently, there was rapid growth in voluntarism (Prochaska, 1990: 357). Philanthropic and voluntary movements of the eighteenth and nineteenth centuries have been given much attention, with debate centring on scope, scale and efficacy of individual institutions and of the movement as a whole (Owens, 1964; Finlayson, 1990; Roberts, 2002; Hilton; McKay, 2011). However, like other fields of study discussed above, the main focus has been land based, overlooking the maritime and environmental dimensions (Kennerley, 2016). This does not mean that the connection of charity and philanthropy to watery worlds has not been investigated. Of particular note are the activities of evangelical naval officers, who were behind the founding of seamen's missions and sailor's homes in major ports (Kverndal, 1986; Press, 1989; Blake, 2014). Other important maritime charity and philanthropic societies established in the period include The Royal Humane Society (Davidson, 2001; Williams, 1996) and the Royal National Lifeboat Institution (Gleeson, 2014). There is still much work to be done to incorporate their contributions into mainstream philanthropic and humanitarian history and to reorient the field to include the maritime sphere.

This chapter qualitatively analyses records of the Shipwrecked Fishermen and Mariner's Benevolent Society (SMS), which are held at their office in Chichester and heretofore have not been used in scholarly research. Other sources discussed include contemporary newspapers and government wreck reports. The chapter aims to highlight the social and material space of the sea and the inhabitants who live beside it, by investigating the ways in which gales and shipwrecks were recorded, transmitted and employed by the Society. It will show how a large national charity, along with its local auxiliaries, experienced and responded to extreme weather that occurred on the coasts of the United Kingdom. The chapter will also seek insight into the response of coastal communities to the advent of the SMS and to the unpredictability of their marine environment.

"The object of which is to afford immediate and effective relief to the survivors": beginnings

The foundation narrative for the SMS is central to their identity. It holds a prominent place on their website (SMS, 2016) and a painting illustrating the moment of inspiration to create the Society takes centre stage in their boardroom in Chichester (Figure 4.1). It is an act of institutional remembering. The painting shows retired physician John Rye of Bath learning about the tragic story of the Clovelly fishing fleet as retired pilot Charles Gee Jones read the news to him in October 1838. Rye was a well-known philanthropist, involved with local charities such as The Humane Society and the Society for the Occasional Relief of the Poor.[5] Shortly afterwards, he contacted Rear Admiral Sir Jahleel Brenton, RN, Lieutenant-Governor of Greenwich Hospital. Rye proposed a fund to assist the widows and orphans of the north Devon gale, as otherwise "there should be no fund to meet the distress it had occasioned".[6] Admiral Brenton championed the idea and drew together some of the most famous names in London's maritime philanthropic circles, including Capt. The Hon. Francis Maude RN, who was involved in an evangelical movement to improve the conditions of seafarers.[7] The great Devon storm had occurred on the night of 28–29 September 1838. In February 1839, the newly formed Shipwrecked Fishermen and Mariners' Benevolent Society held their first meeting in London;[8] a month after that, the Society had gained the patronage of Queen Victoria.[9]

Figure 4.1 John Rye and Charles Gee Jones discussing the loss of the Clovelly fleet, 1839. Courtesy of the Shipwrecked Fishermen and Mariners' Royal Benevolent Society.

From its establishment, the SMS was a voluntary public fund supported nationally by donations from wealthy elites, both named and anonymous (Flew, 2015). Annual subscriptions of 2s 6d were designed to induce fishermen and mariners to manage their own affairs, as a form of self-help (Roberts, 2002: 212). The SMS's first advertisement, sent out to the daily newspapers including the *Times, Morning Chronicle, Morning Herald* and *Morning Advertiser*, emphasised the prominence of extreme weather as their *raison d'être*:

> The <u>tremendous gales</u> which have been experienced during the present winter on the coasts of Great Britain, having been attended with loss of life to immense numbers of our fellow subjects, and with much consequent misery and distress to their families, a Society has been established, the object of which is to afford immediate and effective relief to the survivors...[10]

Thus, the 'objects' of the Society included relief to shipwreck victims by clothing and feeding them; providing medical attention; transporting them home; giving immediate assistance to widows and orphans of fishermen by furnishing them with money for rent; and assisting fishermen who had lost boats and/or nets and who were "left destitute in consequence of extreme distress occasioned by storms".[11] It was soon apparent that they had to limit their financial assistance to members of the Society; however, they continued to help all shipwreck victims, irrespective of their nationality or religion. The Society adopted the use of local honorary agents and auxiliary societies located in individual communities to assist them in their work.

"The hand of charity is paralyzed by want": local and national philanthropy

The SMS was not the only organisation to react to the north Devon gale in September 1838. John Rye was wrong that there was no fund for their relief. On 3 November at Bideford, concerned members of the public met to form the North Devon Shipwreck Society to raise money for the widows and children. They also voted to establish a permanent fund, for "the relief of all mariners, fishermen and their wives and families, who may suffer shipwreck within this district".[12] Thus it had similar objectives as the SMS but on a local level. However, the North Devon Shipwreck Society failed. That November gale was followed by more extreme weather in December–January 1839,[13] and there was no further mention of the North Devon Shipwreck Society in the local papers. The Society limped on through August when they re-quested to join the SMS as an auxiliary society,[14] allowing them access to a greater amount of funding and a national network.

Until the formation of the SMS, most socioeconomic responses to fishing tragedies and shipwrecks occurred at a local level. For fishing communities, in particular, the loss of men and boats affected everyone, as families were inter-married, and brothers, sons, grandsons and nephews worked on each other's boats creating networks of social capital bonding (Pelling and High, 2005: 310). Such networks of people have shared socioeconomic status, identity and "trust-ing and supportive relationships" based on "norms of reciprocal care" that "serve to cement community resilience and to fashion resilience narratives" (Wolf et al., 2010: 49–50). Some communities may have reacted to the risks of extreme weather by resorting to ways of thinking that allowed them to cope, in-cluding cognitive dissonance. This mechanism meant that they recognised the risks they experienced but continued to operate without changing their socio-cultural practices (Festinger, 1957; Wolf et al., 2010: 45). This can be seen in the Scottish insistence on using dangerous open-decked fishing boats (Coull, 1998: 205). However, there is evidence that some communities attempted to mitigate risk from fishing losses by forming local benefit societies for their members. The oldest were founded in Scotland, including the Society of Free Fishermen of Newhaven, near Edinburgh (1572); the Society of Fishermen of Fisherrow (1686); and the Nairn Fishermen's Society (1767).[15]

Many of these societies were able to handle small, local tragedies, but in cases of large-scale disasters, the consequences often went beyond their means. In that case, local elites often established subscriptions for funds to pay for lost boats or nets or to pay rents for the widows and used the press to elicit sympathy and open pockets. A narrativised account of a fishing trag-edy on the north Devon coast on 4 October 1821, fashioned to play on emo-tions, was published by the *Exeter Flying Post* under the by-line "Distressing Catastrophe". It reported the disaster with pathos. More than 40 herring boats wrecked on the rocks during a sudden gale: "The cries of the drown-ing, 35 in number, most of whom have left large families, produced an effect too heart-rending to be adequately expressed". It goes on to say that "The distress occasioned to the families of the unhappy sufferers, who looked forward to a fishery for their entire support, but now, alas! bereft of the means of subsistence, is most afflicting".[16] An identical article appeared on the same day in London's *Public Ledger and Daily Advertiser* under the by-line "Dreadful Shipwreck". Within days shortened versions were disseminated by the *Bristol Mirror* and the *Royal Cornwall Gazette*; all papers printed the names of two clergymen visiting Clovelly at the time of the disaster and who donated £5 for the victims.[17] These stories were published before the Bideford elites and the mayor of Exeter established the formal countywide subscription on 9 October.[18] The subscription announcement was picked up by local and national news outlets a month after the disaster, and eventually £1000 was collected for the widows and children.[19] Contemporary newspapers (see Gardner, 2016) were an important vehicle for disseminating such emotional appeals, allowing societies to reach out to broader social and economic networks to collect much-needed funds. However, each response was reactive, short-lived and could not offer immediate support. The SMS was established as a permanent organisation to counteract such delays.

In the first year of its existence, the SMS was in contact with at least 13 other maritime charitable organisations. Most were local – ranging from the Lincolnshire Coast Shipwreck Association, the Sunderland Society for the Relief of Widows and Orphans of Seamen Belonging to that Port and the Society for the Immediate Relief of Shipwrecked Seamen in Littlehampton.[20] Unfortunately, little research has been done on these societies, so it is difficult to know under what conditions they were founded. The Norfolk Association for Saving the Lives of Shipwrecked Mariners (1823)[21] and the Lincolnshire Coast Shipwreck Association (1827) were formed as life-saving societies by local elites to operate lifeboats and life-saving apparatus (Farr, 1981). The Ipswich Seamen's Shipwreck Society (1826) was established as a benevolent society, supported by subscription and nominally formed to offer financial assistance and pensions to local mariners. They extended their remit to assist shipwreck victims who were not members (Clarke, 1830: 427). However, they, too, showed an interest in uniting with the SMS in 1839.[22] Others attempted to form after the advent of SMS, preferring to go it alone. The attitude of the elite in Carnarvon, a district that saw many shipwrecks, illustrates the conflict between the local and the national in their desire to maintain control. The SMS recognised their concerns but felt that a central fund, because of the large numbers of subscribers, "would be more likely to be adequate to its relief than those of any local Society could be".[23] This point was recognised in Edinburgh after local ship owners, magistrates and church officials tried to establish a shipwreck society in 1840. Because their membership was so small, the "means [were] so inadequate to the claims made upon their sympathies",[24] they joined with SMS.

Some national voluntary societies supported shipwreck victims, albeit that was not their main focus. As Admiral The Hon. Sir George Cockburn, the first Chair of the Society explained, the SMS "in no way interfered with the duties of the Humane Society, or the Seaman's Hospital Society", rather the SMS was "formed in aid and support of these societies, for its functions commenced just at the very point where the duties of the other societies ceased".[25] Two of the most important, like the SMS, still exist – The Humane Society (1774) and the Royal National Lifeboat Institution (1824), originally known as the National Institution for the Preservation of Life from Shipwreck (Coke, 2000; Gleeson, 2014). Like the SMS, both organisations had local representation, either by absorbing like-minded societies already in existence or by spawning local branches. Many SMS members were active in these organisations and cooperated during life-saving emergencies.[26]

Despite the existence of local shipwreck societies and other forms of charity, it was not enough to deal with the pressures from the multitudes of shipwrecks that occurred on British shores, particularly in locations away from the major ports (Table 4.1).

Table 4.1 Sample shipwreck casualties and fatalities on coasts of the British Isles

Year	No. vessels wrecked	Total lost/damaged vessels from 'Stress of Weather'	No. lives lost
1833	651		572
1834	497		578
1835	554		564
1841	630		642
1842	586		608
1850	660	*70+ wrecked in gales between 30 and 31 March	640 *13 vessels 'All' lost *10 vessels '?' unknown
1851	1274	*153+ wrecked in 25–26 September gales	174 *43 vessels 'All' lost *27 vessels '?' unknown
1852	1115		920
1853	832		989
1854	987		1549
1855	Not available		485
1856	837		521
1857	866		539
1858	869		353
1859	1067	606 (343 in Royal Charter storm)	1647
1860	1081	645	3697

*Calculated from Wreck Returns; not supplied in official tabulations.

Compiled from Parliamentary Papers (hereafter PP) (1836) Report from the Select Committee appointed to inquire into the Causes of Shipwrecks, with Minutes of Evidence; PP (1843) First Report from the Select Committee on Shipwrecks, together with minutes of evidence; PP (1852) Admiralty Wreck Register. Return…of the register of wrecks and other casualties to vessels which occurred in the seas and on the shores of the United Kingdom, during the years 1850 and 1851; PP (1853) Register of Wrecks for 1852; PP (1854) Register of Wrecks for 1853; PP (1855) Register of Wrecks for 1854; PP (1865) An abstract of the returns made to the Lords of the Committee of Privy Council for Trade, of wrecks and casualties which occurred on or near the coasts of the United Kingdom.

A general idea of the sheer numbers of vessels wrecked on the British and Irish coasts that are known and recorded can be gained by consulting the official Wreck Returns. However, they are to be used with caution. Statistics were not consistently collected or reported, and they were compiled from disparate sources, including Lloyd's List, Coastguard returns, Lloyd's agents' reports, and the *Shipping and Mercantile Gazette*. The actual numbers of wrecks and lives lost are larger than these figures suggest. The Admiralty Wreck Report for 1852 is particularly problematic in that the compiler did not consistently use the 'lives lost' column, but mentioned deaths in the comments. Also, most lives lost were not tallied, but only listed as 'All' lost with the vessel. These issues point to the problems involved with using such statistical data; however, they can provide an idea of the large numbers of lives lost on the British coastline, as well as the numbers of ships wrecked or stranded, which would result in hundreds of seafarers left without berths far from their home ports.

Some of the wreck returns also indicate how the impacts of extreme weather are not only "unevenly socially distributed" (Morgan, 2015: 47), but unevenly geographically distributed as well. For the returns of 1852, 1853 and 1854, the compiler also enumerated the numbers of wrecks on each coast (Table 4.2).

Thus shipwreck charities would have to be far more active on the east and west coasts, which were supported by the large port towns such as Edinburgh, Newcastle, Hartlepool, Hull, Great Yarmouth and London in the East and Liverpool, Milford Haven, Cardiff and Bristol in the West. The SMS was concerned to reach out to more remote locations. As Admiral Brenton said, "These calamities most frequently occur in the lonely and desolate parts of the coast... upon the open and thinly inhabited parts, where shipwrecks so frequently abound, however strong the inclination to afford relief, the hand of charity is paralyzed by want".[27] Local honorary agents were appointed, and auxiliary branches of the SMS were established to

Table 4.2 Wrecks by geographical areas

	1852	1853	1854
East Coast	464	253	350
South Coast	158	76	38
West Coast	235	180	164
Irish Coast	128	81	68
Isles of Scilly	5	6	5
Orkney and Shetlands	18	3	19
Channel Islands	–	11	9
Isle of Man	18	12	5
'Surrounding Seas'	80	260	331

Compiled from PP (1853) Register of Wrecks for 1852; PP (1854) Register of Wrecks for 1853; PP (1855) Register of Wrecks for 1854.

respond quickly to need and to connect with the Central Society when more funds were required. Indeed, within a year of its launch, over 57 auxiliary branches and 89 honorary agents represented the SMS around the British and Irish coastlines.[28] Many honorary agents were already acting agents for Lloyd's insurance, such as Edward Chatterton of Rye and Thomas Gillaume of Shoreham.[29] Others were Coastguardsmen, naval officers, rectors, ship-builders and local businessmen. By the end of the nineteenth century, there were over 1,000 auxiliaries and honorary agents, including those in foreign ports such as Antwerp, Dieppe, Genoa, Malta, Newfoundland, Shanghai and Yokohama, among others, and over 522,375 people had been assisted since 1839.[30]

"To … alleviate and soothe such poignant sorrows": extreme weather and funding

The SMS records offer a useful source to study the relationship between extreme weather events and a philanthropic organisation. The extant records include minute books, proceedings, annual reports and their *Shipwrecked Mariner's Magazine*, published from 1854 to 1895. Unfortunately, the petitions and correspondence books have not survived. Neither have the photographs of corpses taken as they were interred by the Sheffield auxiliary of the SMS in 1873 "as a source of consolation to the relatives of the deceased to know that they had been decently buried".[31] However, those records that are available show how the SMS made use of extreme weather narratives to further their funding drives. The records also reveal how weather affected operations. Although shipwrecks are caused by many factors, such as defective construction, inadequate equipment, poor repair, improper or excessive loading of cargo and incompetent masters and officers,[32] the weather underpinned the rationale of the SMS. Extreme weather narratives were used to draw on the sympathy of readers so they would donate, or, if fishermen or mariners, subscribe to the Society. The SMS drew on the same language and discourse for their speeches as was used by elites and news editors in local subscription drives. Indeed, the shipwreck narrative was a reoccurring trope, using classic Victorian melodramatic language and instituting pathos in a 'performance' in public meetings to elicit compassion. The speeches were then reprinted in newspapers and annual reports for wider consumption, in the hope of gaining even more donors.[33] Sir Robert Peel's speech at the second festival dinner of the SMS in London in April 1840 is a case in point. Invited as a special guest at a time when he led the opposition in Parliament, he drew on the sympathies of the audience by dramatizing the dangers mariners and fishermen faced:

> All of you will bear witness that, in the interior of this country, when the tempest is raging, the first congratulation of those who are not exposed to its horrors is, "Thank God we are safe;" and then the expression that

proceeds from every lip is, "What a dreadful night this must be at sea!... How thankful ought we be that we are secure, and protected from the dangers of the ocean!"

He used sympathetic exchange to induce the listener to imagine the experience, to recognise their common humanity, but he also reminded them of their distance from dangers experienced by men on whose labour the nation depended. A thankful audience, Peel hoped, was one that felt 'distressed' by the sufferings of the mariners and thus would "at a very small and cheap sacrifice, purchase a consolation of the most precious nature".[34] Peel's speech also reminds us of the importance of the sea in British affairs.

Similar emotional and melodramatic phrasing is evident in most of the annual reports. In 1841, just after the loss of three fishing boats in Mount's Bay, the report stated that:

> the tremendous storms of last winter were productive of numerous wrecks, and of the destitution and misery which are ever their attendants...When the painful accounts of those accidents, and the still more distressing details of the widow and orphan having to mourn the stay and support which had been taken away from them, were daily, almost hourly, brought under the notice of the Central Board, they then felt the inestimable advantage of a Society which could, to some degree, alleviate and soothe such poignant sorrows.[35]

The use of the common linguistic tropes such as 'misery', 'painful', 'distressing', 'poignant' and 'sorrows' brought home to readers the need for compassion. The language connected the powerful effects of the gales to the personal. As Julie-Marie Strange argues, the reader's "capacity for compassion rests partially on imagining physiological sensation: the reader must feel feeling" (Mason, 2007; Strange 2011: 247). To give donors a sense of moral gratification, the Society reports also enumerated the numbers of victims who had been assisted through their donations. In 1841, for instance, the Society had relieved 126 widows, 87 aged parents, 1,007 shipwrecked persons, and 205 families "who were left destitute in consequence of extreme distress occasioned by storms".[36]

"The views of the society are confined to sudden storms": cases for relief

The centrality of extreme weather to the SMS is also evidenced in their rules that relief should only be afforded to "widows and orphans of fishermen; and of mariners, members of the Society, who lose their lives by storms and shipwreck".[37] This imperative is demonstrated in petitions for assistance that were denied. John Thomas, a dredgerman from Gillingham, Kent, lost his vessel by grounding. As much as the Committee wanted to help, they sent a reply "that the Views of the Society are confined to sudden Storms & not from accidental occurrences or getting aground in difficult or dangerous passages".[38] Likewise, a Hastings fisherman named Apps was killed while fishing in the Channel. The local agent's request for relief was refused, as Apps had died from a collision with a Government steamer, rather than from a storm, and it was the Committee's contention that the widow would receive Government compensation.[39]

Although weather was the qualifying factor in the issuance of relief, the SMS was also concerned about the morality of the recipients, as were other philanthropic societies of the time. They believed that the purpose of assistance was to allow the fishermen or mariners, if they survived, to return to vessels so they could "get a livelihood". The Central Committee required each case to be accompanied with 'particulars', including the details of the accident, the number of dependents, cost of rent, and information on their morality. Indeed, petitions for assistance were often verified by inquiries to local officials, to check on the weather conditions of the accident and the person's morality and to prevent fraud. Owen Sullivan's petition, whereby he sought redress for the loss of his boat in a gale off Bere Island, Ireland, was corroborated by letters from the parish priest, a justice of the peace and the local Lloyd's agent.[40] Sometimes the Committee offered assistance to non-subscribers of particularly high merit. The widow of drowned fisherman James Pengally, of Looe, in Cornwall, was awarded £8 because of the good character "of her late husband, who had saved at different times by jumping into the sea 20 lives; and had brought up and put into the world seven orphans, and lost his life attempting to save the property of others".[41] Thus, assistance was only given to those "as may lose their lives in their hardy and useful, but often dangerous, occupation"[42] as long as those lives were lost from storms.

"Unaccountable indifference towards the purposes of this charity": voices of coastal communities

Absent from these accounts, however, are stories that particularize how the local inhabitants perceived their experiences and encounters with dangerous climatic conditions. The voices of individual victims are for the most part silent in SMS records – the petitions having been lost or destroyed long ago. However, we can glimpse how people responded – or did not – to risks in their environment and to a society formed to aid them in coping with disasters, albeit mediated through the dominant discourse of the elites of the SMS. Some fishing communities were reluctant to entertain any social adaptation to environmental risks that reached beyond their own networks of social capital, and they "showed a distinct mistrust of linking ties" (Pelling and High, 2005: 310). The records frequently allude to a frustration that fishermen were not subscribing, despite the fact that the Society was initially established with them in mind. In 1841, the numbers of fishermen subscribers had increased, but there was still concern on the part of the SMS. The secretary, Edward West, wrote that "the Committee are pleased to find, that the unaccountable indifference towards the purposes of this Charity, which has hitherto manifested itself amongst the Fishermen of several populous districts, it is beginning to give way to more correct feelings and better judgement".[43]

The fishermen's reticence was blamed directly on what was assumed to be their lack of knowledge about the objectives of the society. This opinion was echoed in some of the outward localities. In Chelmsford, frustration was voiced when a meeting of the Leigh Auxiliary Branch had to adjourn early, because of "the thin attendance of the fishermen, and others whose especial benefit the society has been formed". The *Essex Standard* went on to chastise them: "Such conduct is but little calculated to encourage those who are labouring for the good of their poorer neighbours". It suggested that the fishermen, "from selfish miscalculation, or thankless indifference...may, in the hour of their calamity, keenly regret their short-sightedness and gratitude".[44] In 1848, the Committee stated that, "the Fishermen, as a body, have not manifested as great a zeal and anxiety as Mariners in supporting the Society, which was established originally for their relief".[45] Indeed, the Society had not planned to give mariners the same level of benefit as the fishermen and their families. Initially, mariners were only assisted during shipwreck and given transportation to their homes. Fishermen, more than almost any other occupation, were reliant on the environment. Fishing was, and still is, considered to be the riskiest of occupations due to its vulnerability to storms and severe weather (Roberts, 2010; Byard, 2013).

Why were many fishermen so reluctant to join SMS when it offered protection to their families if they were drowned during storms? Why did so many of them fail to adopt it as a coping strategy to deal with risks of their environment? It could be argued, as do Wolf et al. (2010), that bonding mechanisms "give the narratives of independence and resilience a permanence that prevents proactive anticipatory and long-term adaptation" to risk (50). Unfortunately, the records do not really give a full indication if such bonding mechanisms were behind the fishermen's decision making, nor do they give enough detail to answer these questions. From the Committee's viewpoint, their reluctance was partially economic. Two shillings 6d was a trifling amount for a subscription, and while the Committee recognised that fishermen might belong to other benefit societies, they did not seem to take into account that even a small annual subscription could be onerous. Many benefit societies required monthly membership fees, such as the Fishermen's Refuge on the Yorkshire Coast, while others required fees of one shilling per week for nine months, such as the United Fishermen's Society of Brighton (Bruce, 1835: 76; Theakston, 1845: 95). The overall financial picture of many fishermen was poor; the costs they had to bear for boats, nets, taxes and tithes could be burdensome. Their profit from fishing was as unpredictable as the shoals of herring and mackerel they relied on. Many could not even afford their own boats, but rented them (Coull, 1998: 212). However, costs were just one factor.

The added dynamic of national versus local control must also be considered. The SMS was a national, London-based charity and was viewed suspiciously by some. We have seen how the elites of Carnarvon were not interested in forming an auxiliary society because they wanted local control. The same could be said of locations such as Greenock, where "a prejudice exists there against the establishment of a branch...on account of the concentration of funds in London".[46] Social status and ties of social capital, too, might have been a factor. Even though the Society from the outset relied on local honorary agents, the agents were usually from the middle ranks and thus outside of the networks of social bonding. They were local, but they were not members of the tight-knit fishing communities who would rather take care of their own (Thompson et al., 1983). As Alan Kidd (1996) has argued, "charitable gifts to the poor in an unequal society have... masked the presumption of benefactors, and placed the act outside of the sphere of mutual ties (as in family relationships)" (86–87). Local disasters, too, however, had the additional effect of strengthening bonding ties rather than inducing individuals to form greater interpersonal networks across social boundaries (Pelling and High, 2005: 310).

The response of mariners was more positive, particularly those from the active ports of northern England such as Sunderland and Newcastle, which saw the greatest numbers of shipwrecks (Larn and Larn, 1995–98) and who did not operate from such closely knit communities as the fishermen. Some captains and managing owners insisted that their crews become subscribers and even helped them to pay the annual 2s 6d.[47] In 1842, the Committee announced that "so anxious have the seamen, in the north of England, been to enter their names in the Books of the Society... that they have visited the Office in London, in considerable numbers, for the purpose of subscribing".[48] Just as coastal communities reacted to extreme weather by forming shipwreck societies, so too did frequent gales provoke some mariners to quickly join the SMS.

Although many individual fishermen and mariners chose not to join the SMS, there were thousands who did. The records of the SMS include names of those who drowned or were assisted. They also contain evidence of how individuals and families were affected by extreme weather, whether from the loss of boats, nets or lives. Additional in-depth research on these individuals can be a valuable angle from which to construct the social makeup of coastal communities and their bonds of social capital and to assess their sociocultural responses to place, risk and extreme weather. By so doing, valuable narratives of human interaction to environmental challenges can be uncovered.

"Many melancholy cases of wreck": the spatial impact of extreme weather

Because the SMS was a national organisation and was established through-out many different geographies, the records are useful for spatial analysis. Agents represented the Society in large shipping ports such as Liverpool, Newcastle, London and Cardiff, as well as small fishing ports and villages such as Brixham, Aldeburgh, Blyth and Stonehaven. Some ports were mixed economy. Although the larger shipping ports had other active local socie-ties, they also provided many cases for the SMS. Agent James H. Palmer of Great Yarmouth in Norfolk was one of the most active. After establishing the Yarmouth branch in August 1839, he gave relief to the first shipwreck victims in September when the crews of the *James* of Whitby and *Triad* of Sunderland had wrecked on Scroby Sand. In October, he relieved the crew of the *Cleveland* of Hull, which also found grief on Scroby Sand. The men had been provided with shoes, provisions and a free pass to Hull. In December, Palmer "forwarded the wife & child of the Master of the Broth-ers wrecked at Horsey, to Boston".[49] Indeed, cases from Yarmouth seemed to occur monthly, which indicates the sheer volume of shipping traffic, the number of navigational hazards in that area, and a proactive SMS honorary agent. Other locations had infrequent cause to provide relief. The Brighton auxiliary, for example, dealt with cases of both shipwrecked fishermen and mariners in its first year but in its 1842 general meeting reported that "there had been very little demand on the funds of the Society [as] there had been... very little of calamity on the other".[50] Most auxiliaries reported no activity at all. Of course, the need for relief depended on geography and the anoma-lies of annual climatic patterns.

Although the Central SMS was busy assisting shipwreck victims from the usual storms and gales, particularly those in the autumn and winter months, they noted extreme events because of the toll they took on their funds and the stress and strain on local honorary agents who were in the front line. Three examples offer a means to assess the differential spatial impact of extreme weather: the Mount's Bay disaster of 1840, which affected a sin-gle fishing fleet, the widespread 'hurricane' of 1843 and the 'Royal Charter' storm of 1859, which affected fishing and shipping alike.

The SMS had been in existence for only three months and was barely op-erational when it received word of a fishing disaster in Cornwall on 8 May 1839. Three Mount's Bay mackerel fishing boats were lost during a 'furious storm'.[51] Twenty-one men were drowned, "leaving 12 widows, 7 aged par-ents" and 35 children.[52] A local Penzance committee began a public sub-scription. Messrs. Bolitho, of Penzance and Fenchurch Street, prominent shipping insurance brokers and merchants, approached SMS for additional funds. The SMS Committee responded by holding a special meeting and then donated £30, even though "the Society is not yet in a mature state for making grants of relief generally". They requested further information about the widows, to see if they could manage to grant additional relief. £50 more was sent to Penzance.[53] In the end, together with the two SMS contributions, which were the largest single donations, £672 3s 8d was collected for the families.[54]

The Society's first quarterly report listed the names of those assisted, making the tragedy personal: William Oats, aged 67, and his wife, aged 61, lost their son, "who was of great assistance to them". Alice Kelynack, aged 69, lost her son, as well as her interest in one of the wrecked boats. She had a widowed daughter with four small children living with her. Jan Roufignac lost her husband and a son, as well as all their nets. She was in delicate health yet had to care for five children still at home. Susan Hitchens, already in debt for £4.10 for rent and nets, lost her husband and all the nets. She also had to care for her "idiot sister".[55] The *Royal Cornwall Gazette* reported that one of the boats, the *Bounty* of Mousehole, had a crew that comprised the master, his two sons, a brother and two cousins. All were lost. In this case, the storm affected a relatively small geographical area, but it was devastating for the Mount's Bay communities. The St. Ives mackerel fleet was also out that night, but all were able to make it to shore safely, although there was much damage and loss of expensive nets. The correspondent ended his report with a standard phrase marker: "in every way this disaster is more severe than any similar calamity which has visited our fishermen for many years".[56]

The second case of extreme weather affected a much larger area and a greater number of people; fishing boats and merchant ships alike were wrecked. On 13 January 1843, a gale, which news correspondents soon termed a 'hurricane', hit the British Isles. Using previous storms as a benchmark, the *Morning Post* claimed it was "the severest that has occurred for many winters". For Plymouth, the gale's "wind, rain, thunder and lightning, [was] unequalled in severity since the fatal gale of 1824". The Dublin correspondent stated that "we scarcely remember greater severity at this season. The mercury in the barometer fell lower yesterday than has been observed since the eventful storm of January 1839".[57] The reports summarised just a few of the hundreds of shipwrecks that night. The lifeboats and Coastguard were busy, bringing crews and passengers to safety. But whether seamen and fishermen were drowned, or were shipwrecked, the SMS honorary agents also had their hands full.

Letters from honorary agents requesting additional funds for victims of the gale flooded into the Central Society. County Down in Ireland had been particularly hard hit. William Murrow, the agent for Annalong, reported that 27 fishermen had been lost, leaving ten widows, 42 orphans, and "several Widowed Mothers... in great distress".[58] The following week, the Committee received a letter from Capt. Janns, agent for Dublin. He reported the total loss of 73 fishermen from Newcastle and Annalong, who left 49 widows, 87 children, 40 aged parents and other dependent relatives. A local subscription had been formed for the widows. The Board voted to send £100 to him to add to the local subscription. It was not over. Capt. Janns wrote again on 3 February, informing them of the loss of 18 fishermen from Arranmore and mainland County Donegal. Eight more widows, with 28 children and 12 'aged' parents were added to the list. Requests came in from other parts of Great Britain as well.[59] The Society estimated that on 13 January, at least 180 British merchant ships wrecked on the coasts of the United Kingdom, Ireland and France, and that "433 persons perished therein".[60] The government, however, estimated that over 240 ships were lost over three days, taking 500 lives (Williams, 2005: 193). These numbers did not include the loss of fishermen.

The burden on SMS funds was immense; they could not relieve everyone who petitioned for assistance. They had to limit payments to widows whose husbands had been members of the Society and who had drowned in the gales. The desperation of the Committee for more funding is evidenced by the letters that the secretary sent – one to Earl de Grey, the Lord Lieutenant of Ireland, a Vice President of the Society; one to the Corporation of London; and several to the Inns of Court. The letter to Earl de Grey was to "detail the many heavy cases recently relieved in Ireland, and the little support which this Charity has received from that Country, and requesting his Lordship's influence upon the subject".[61] In this case, neither the localities nor the Central SMS found it easy to deal with the immense losses.

The last example comes from the 'Royal Charter' storm, although the storms discussed in this chapter are by no means the only examples of extreme weather in the Society's records. However, the 'Royal Charter' storm is perhaps one of the best known, barring the 1703 Great Storm immortalized by Daniel Defoe (1704, 2003). Named after an iron steamship lost on the rocks off Anglesey on 25 October 1859, the gale had caused wrecks all around the coasts of the United Kingdom. Public attention, however, focused on the one ship. The story of the *Royal Charter*'s loss is horrific: 450 passengers and crew died that night. Only 50 were rescued. As Krista Lysack (n.d.) argues,

> this wide-ranging storm… required a focal point in the papers, and the ship's distress served as one for the growing and corresponding media storm: a means by which to transform for readers a robust weather occurrence into a narrative about human fear, heroism, and in the end, death for most.

Charles Dickens, too, wrote poignantly about the wreck in his periodical *All the Year Round*. He reprinted the article in *The Uncommercial Traveller* (1860: 3–22), along with correspondence inspired by the original article.[62]

For the SMS, however, the wreck of the *Royal Charter*, as with all of the other wrecks and groundings from the gale, required local action. The minutes of the Society for this period unfortunately lack details. By 1859–1860, the secretary only recorded changes to honorary agents, occasional correspondence on possible lifesaving medal recipients, and the Society's financial report.[63] However, he also wrote a list of all of the beneficiaries awarded funds that week. It is difficult to tell which fishermen and mariners were victims of the 'Royal Charter' storm as opposed to other storms, as dates of their accidents were not given, but at the 4 November meeting, the first held since the gale, the Committee accepted 274 applicants that week alone.[64] The Committee had met monthly through the summer but began to meet weekly after November and continued to do so through that winter. Each Friday they accepted, on average, over 100 applications.[65] They did not record the rejections.

The lead article in the *Shipwrecked Mariner's Magazine* for January 1860 was obviously inspired by the 'Royal Charter' gale. It underscored the danger of extreme weather, the perils offered by coastal geography and the inability of humans to control nature. The "material and tangible reality of water worlds" (Anderson and Peters, 2014: 15) is made manifest:

> His ship may be of unequalled strength and managed with matchless skill; he may have the most excellent charts, the most correct knowledge of his position, and be familiar with every headland and rock, with every creek and shelter, with the currents and soundings…when suddenly, without any indication of warning, a tempest may arise such as he never witnessed before, which will dash his mighty vessel to pieces on the shore, and scatter the lifeless bodies of himself and his hapless companions in every direction…

One hundred and ninety-five vessels were wrecked in the gale (official statistics later claimed 343 wrecks), "with the loss of 684 of our fellow-creatures, including those of the 'Royal Charter'". The Society paid out an average of £350 a week and over £10,000 for the year 1859, "whilst applications from sufferers of the late storms continue to pour in, so that the Committee have compelled to sell £1000 stock out of their little funded property, which is set apart to give small annual grants to help pay the rent of cottages of about 600 widows and orphans (Figure 4.2)".[66]

Conclusion

Environmental risks and losses were frequently borne most heavily by the fishing and seafaring communities. The impact of extreme weather was unevenly distributed – socially, economically and geographically. This is shown by the individuals assisted by the SMS and by the differing level of activity of auxiliary societies as they grappled with the unpredictability of weather. As well, these same fishermen and mariners were placed under economic and political pressures from fisheries policies and labour law. Many were forced to go to sea when they knew the conditions were dangerous. If they refused on the grounds of believing their vessel was unseaworthy, they could be incarcerated for up to three years (Wilcox, 2008: 181; Jones, 2006: 13–14).

Figure 4.2 The *Shipwrecked Mariner* Magazine. Courtesy of the Shipwrecked Fishermen and Mariners' Royal Benevolent Society.

From its inception, the Shipwrecked Mariners' Society concerned itself
with uncontrollable environmental threats to fishermen and mariners. At
the forefront of the Central Committee's decisions as to whether or not they
would grant assistance was the role of weather in the resultant loss, particularly seasonal gales and extreme weather. Collisions and navigational errors
were thought to be controllable, and thus compensation was expected to
come from the transgressors. Later in the century, after the SMS gained
a stronger financial footing, they relaxed the requirement that losses had

to be caused by storms and offered assistance to fishermen's and mariner's widows whose husbands had died of cholera, yellow fever and consumption. However, severe weather remained at the centre of their *raison d'etre*, connecting climate, the marine environment and coastal populations to the more land-based views of charity and philanthropy.

Acknowledgements

I would like to thank Commodore Malcolm Williams RN of the Shipwrecked Mariners' Society, for unlimited access to the Society's archives and for the able assistance of his staff. A note of thanks also goes to Georgina Endfield and Lucy Veale for inviting me to participate in this project.

Notes

1 *North Devon Journal*, 1 November 1838.
2 The numbers of children were not stated.
3 *North Devon Journal*, 1 November 1838.
4 For the purpose of this chapter, 'extreme weather' is identified by phrase markers used in reports: storms, the scale of which are perceived by those experiencing them as being "not remembered for a great many years", or "exceeded in violence to any that had been known in the memory of man" and have caused either a tremendous loss of life or property. Although this language is used in the media to introduce drama to weather events, the use of generational language as such is common in collective memory and has even been used to establish legal precedent of use and ownership of manorial rights (Wood, 2013).
5 *Bath Chronicle and Weekly Gazette*, 23 June 1831, 5 April 1832, 15 January 1835.
6 SMS First Annual Report, 1839.
7 *Shipwrecked Mariners Magazine* 1887: 65.
8 SMS Minute Book I, 21 February 1839: 1.
9 SMS Minute Book, I, 26 March 1839: 7.
10 SMS Minute Book I, 4 March 1839: 4.
11 SMS Second Annual Report, 30 April 1840.
12 *North Devon Journal*, 8 November 1839.
13 *North Devon Journal*, 27 December 1838; 17 January 1849.
14 SMS Minute Book I, 7 & 8 August 1839: 133.
15 A study of these fishing societies is needed.
16 *Exeter Flying Post*, 11 October 1821; *Public Ledger and Daily Advertiser*, 11 October 1821.
17 *Bristol Mirror*, 13 October 1821; *Royal Cornwall Gazette*, 13 October 1821.
18 *Exeter Flying Post*, 1 November 1821.
19 *Morning Post*, 5 November 1821; *Hereford Journal*, 7 November 1821.
20 SMS Minute Book I, 1839–1840.
21 *Norfolk Chronicle*, 29 November 1823.
22 SMS Minute Book I, 17 August 1839: 138.
23 SMS Minute Book I, 12 December 1839: 195.
24 *John O'Groat Journal*, 14 February 1840.
25 *London Standard*, 9 May 1840.
26 *Brighton Gazette*, 19 November 1840.
27 SMS First Annual Report, 21 February 1839.
28 SMS First Annual Report, 30 April 1840.

29 SMS Minute Book I, 2 October 1839: 154.
30 SMS Annual Report, 1899.
31 *Sheffield Daily Telegraph*, 17 April 1873.
32 PP 1836: 5.
33 Julie-Marie Strange analyses this use of "sentiment as a tool to motivate practical compassion" by a philanthropic society moved to assist individuals who were not as acceptable to society, homeless men (2011).
34 SMS Third Quarterly Statement, 1 April 1840.
35 SMS Annual Report, 1841.
36 SMS Second Annual Report, 1841.
37 SMS First Annual Report, 1840.
38 SMS Minute Book I, 20 June 1839: 35, 102.
39 SMS Minute Book II, 20 January 1843: 233.
40 SMS Minute Book I, 16 November 1839: 176.
41 SMS Seventh Annual Report, 1845–1846.
42 SMS Minute Book I, 10 June 1839: 94.
43 SMS Second Annual Report, 1841.
44 *Essex Standard*, 22 May 1840.
45 SMS Ninth Annual Report, 1848.
46 SMS Minute Book I, 26 December 139: 208.
47 SMS Minute Book I, 27 August 1839: 141.
48 SMS Third Annual Report, 6 May, 1842.
49 SMS Minute Book I, 26 August 1839: 140; 28 September 1839: 153; 29 October 1839: 166; 11 December 1839: 195; 31 October 1839: 170.
50 *Brighton Gazette*, 3 February 1842.
51 *Royal Cornwall Gazette*, 17 May 1839.
52 SMS Second Festival Dinner Report, *Norfolk Chronicle*, 4 April 1840.
53 SMS Minute Book I, 23 May 1839: 82; 17 June 1839: 99; 20 June 1839: 101.
54 *Royal Cornwall Gazette*, 16 August 1839.
55 SMS First Quarterly Statement, 31 October 1839.
56 *Royal Cornwall Gazette*, 17 May 1839.
57 *Morning Post*, 16 January 1843.
58 SMS Minute Book II, 20 January 1843: 230.
59 SMS Minute Book II, 27 January 1843: 237.
60 SMS Annual Report, 1843.
61 SMS Minute Book II, 20 January 1843: 235; 3 February 1843: 140–141.
62 For additional examples of material history from the Royal Charter wreck, see www.peoplescollection.wales/collections/377940 and http://heritageofwalesnews.blogspot.co.uk/2013/12/royal-charter-voyage-journal-of.html (accessed 15 June 2016).
63 It was typical that early minute books were "more charming and revealing", while later minute books were terser and less informative. Checkland suggests this occurred as societies moved to professional secretaries. This certainly seems to be the case for SMS (Checkland, 1980: 8–9).
64 SMS Minute Book VI, 4 November 1859: 111–117.
65 SMS Minute Book VI, 4 November 1859–25 March 1860.
66 *Shipwrecked Mariners Magazine*, 1860: 1, 16.

References

n.a. (1860) Autumnal storms and shipwrecks. *Shipwrecked Mariners Magazine*, VII (XXV, January): 1–16.

Anderson K (2005) *Predicting the Weather: Victorians and the Science of Meteorology*. Chicago, IL and London: University of Chicago Press.

Anderson J and Peters K (2014) *Water Worlds: Human Geographies of the Ocean.* Farnham: Ashgate.

Blake R (2014) *Religion in the British Navy, 1815–1879: Piety and Professionalism.* Woodbridge: Boydell Press.

Bolster WJ (2006) Opportunities in marine environmental history. *Environmental History*, 11: 567–597.

Bolster WJ (2008) Putting the ocean in Atlantic history: maritime communities and marine ecology in the Northwest Atlantic. *American Historical Review*, 113: 19–47.

Bruce J (1835) *The History of Brighton, with the Latest Improvements, to 1835.* Fourth edition. London: J. Bruce. Available at www.googlebooks.com (accessed 25 June 2016).

Burton J (1986) Robert FitzRoy and the early history of the Meteorological Office. *The British Journal for the History of Science*, 19: 147–176.

Byard RW (2013) Commercial fishing industry deaths-forensic issues. *Journal of Forensic and Legal Medicine*, 20: 129–132.

Catchpole AJW (1995) Hudson's Bay Company ships' log-books as sources of sea ice data, 1751–1870. In Bradley RS and Jones PD (eds.) *Climate Since AD 1500.* London: Routledge: 17–39.

Checkland O (1980) *Philanthropy in Victorian Scotland: Social Welfare and the Voluntary Principle.* Edinburgh: John Donald.

Clarke GR (1830) *The History and Description of the Town and Borough of Ipswich.* Ipswich: S. Piper. Available at www.googlebooks.com (accessed 15 May 2016).

Coke D (2000) *Saved from a Watery Grave.* London: The Royal Humane Society.

Corbin A (1994) *The Lure of the Sea: The Discovery of the Seaside in the Western World, 1750–1840.* Translated by Jocelyn Phelps. Berkeley and Los Angeles: University of California Press.

Coull, James R. (1998) National, regional, and local divergences within the Scottish fishing industry since 1800. *International Journal of Maritime History*, 10: 201–218.

Davidson L (2001) *Raising Up Humanity: A Cultural History of Resuscitation and the Royal Humane Society of London, 1774–1808.* Unpublished PhD thesis, University of York.

Defoe D (1704, 2003) *The Storm.* Edited with an Introduction and Notes by Richard Hamblyn. London: Penguin Books.

Dickens C (1860) *The Uncommercial Traveller.* London: Chapman and Hall. Available at https://archive.org/details/uncommercialtrav00dick (accessed 23 March 2016).

Farr G (1981) *The Lincolnshire Coast Shipwreck Association, 1827–1864.* Bristol: Graham Farr.

Festinger L (1957) *A Theory of Cognitive Dissonance.* Stanford, CA: Stanford University Press.

Finlayson G (1990) *Citizen, State and Social Welfare in Britain, 1830–1990.* Oxford: Oxford University Press.

Flew S (2015) Unveiling the anonymous philanthropist: charity in the nineteenth century. *Journal of Victorian Culture*, 20 (1): 20–33.

Gardner VE (2016) *The Business of News in England, 1760–1820.* Basingstoke: Palgrave Macmillan.

Gleeson J (2014) *The Lifeboat Baronet: Launching the RNLI.* Stroud: The History Press.

Hilton M and McKay J (eds.) (2011) *The Ages of Voluntarism: How We Got to the Big Society.* Oxford: Oxford University Press.

Holm P, Smith T and Starkey DJ (eds.) (2001) *The Exploited Seas: New Directions for Marine Environmental History*. St. John's: International Maritime Economic History Association/Census of Marine Life.

Jones N (2006) *The Plimsoll Sensation: The Great Campaign to Save Lives at Sea*. London: Abacus.

Kennerley A (2016) Welfare in British merchant seafaring. *International Journal of Maritime History*, 28 (2): 356–375.

Kidd AJ (1996) Philanthropy and the 'Social History Paradigm'. *Social History*, 21: 180–192.

Küttel M, Xoplaki E, Gallego D and et al. (2010) The importance of ship log data: reconstructing North Atlantic, European, and Mediterranean sea level pressure fields back to 1750. *Climate Dynamics*, 34 (7/8): 1115–1128.

Kverndal R (1996) *Seamen's Missions: Their Origin and Early Growth*. Pasadena: William Cary Library.

Land I (2007) Review essay—Tidal waves: The new coastal history. *Journal of Social History*, 40: 731–743.

Lambert D, Martins L and Ogborn M (2006) Currents, visions and voyages: historical geographies of the sea. *Journal of Historical Geography*, 32, 479–493.

Larn R and Larn B (1995–98) *Shipwreck Index of the British Isles*, 5 vols. London: Lloyd's Register.

Lysack K (n.d.) The Royal Charter Storm, 25–26 October 1859, BRANCH: Britain, Representation and Nineteenth-Century History. Ed. Dino Franco Felluga. Extension of Romanticism and Victorianism on the Net. Available at www.branchcollective.org/?ps_articles=krista-lysack-the-royal-charter-storm-25-26-october-1859 (accessed 23 August 2015).

Mason E (2007) Feeling Dickensian feeling. In Brown N (ed.) *Rethinking Victorian Sentimentality* 19: *Interdisciplinary Studies in the Long Nineteenth Century*, no. 4. Available at www.19.bbk.ac.uk/articles/abstract/10.16995/ntn.454/ (accessed 7 June 2016).

Mathieson C. (2016) *Sea Narratives: Cultural Responses to the Sea, 1600 to the Present*. London: Palgrave Macmillan.

Morgan JE (2015) Understanding flooding in early modern England. *Journal of Historical Geography*, 50: 37–50.

O'Hara G (2009) The sea is swinging into view: Modern British maritime history in a globalised world. *English Historical Review*, 124: 1109–1134.

Owen D (1964) *English Philanthropy, 1660–1960*. Cambridge, MA: Harvard University Press.

Pelling M and High C (2005) Understanding adaptation: What can social capital offer assessments of adaptive capacity? *Global Environmental Change*, 15: 308–319.

Press J (1989) Philanthropy and the British shipping industry, 1815–1860. *International Journal of Maritime History*, 1 (1): 107–127.

Prochaska F (1990) Philanthropy. In Thomson FML (ed.) *Cambridge Social History of Britain 1750–1950*, vol. 3. Cambridge: Cambridge University Press: 357–393.

Roberts F (2002) *The Social Conscience of the Early Victorians*. Stanford, CA: Stanford University Press.

Roberts SE (2010) Britain's most hazardous occupation: commercial fishing. *Accident Analysis and Prevention*, 42: 44–49.

Rozwadowski HM (2013) The promise of ocean history for environmental history. *Journal of American History*, 100 (1): 136–139.

Shipwrecked Fishermen and Mariners' Royal Benevolent Society website. Available from http://shipwreckedmariners.org.uk/who-we-are/our-history/ (accessed 19 May 2016).

Strange J-M (2011) Tramp: sentiment and the homeless man in the late-Victorian and Edwardian city. *Journal of Victorian Culture*, 16 (2): 242–258.

Theakston SW (1845) *Theakston's Guide to Scarborough: Comprising a Brief Sketch of the Antiquities, Natural Protections and Romantic Scenery of the Town and Neighbourhood*. Third Edition. Scarborough: n.p. Available at www.googlebooks. com (accessed 24 June 2016).

Thompson P, Wailey T and Lummis T (1983) *Living the Fishing*. London: Routledge.

Wheeler D (2014) Hubert Lamb's 'treasure trove': ships' logbooks in climate Research. *Weather*, 69 (5): 133–139.

Wheeler D and García-Herrera R (2008) Ships' logbooks in climatological research: Reflections and Prospects. *Annals of the New York Academy of Sciences*, 1146 (1): 1–15.

Wilcox M (2008) The role of apprenticed labour in the English fisheries, 1850–1914. In Scholl L and Williams DM (eds.) *Crisis and Transition: Maritime Sectors in the North Sea Region, 1790–1940*. Bremerhaven: Hauschild: 177–188.

Williams CD (1996) "The Luxury of Doing Good": benevolence, sensibility, and the Royal Humane Society. In Porter R and Roberts MR (eds.) *Pleasure in the Eighteenth Century*. London: Macmillan: 77–107.

Williams DM (2005) Advances in safety at sea in the nineteenth century: the British experience and influence. In Starkey DJ and Hahn-Pedersen M (eds.) *Bridging Troubled Waters: Conflict and Cooperation in the North Sea Region since 1550*. Esbjerg: Fiskeri-og Søfartsmuseets: 177–197.

Williams DM (2010) Humankind and the sea: the changing relationship since the mid-eighteenth century. *International Journal of Maritime History*, 23 (1): 1–14.

Wolf J, Adger NW, Lorenzoni I, Abrahamson V and Raine R (2010) Social capital, individual responses to heat waves and climate change adaptation: an empirical study of two UK cities. *Global Environmental Change*, 20: 44–52.

Wood A (2013) *The Memory of the People: Custom and Popular Senses of the Past in Early Modern England*. Cambridge: Cambridge University Press.

5 The temporal memory of major hurricanes

Cary J. Mock

Introduction

Tropical cyclones are a significant, life-threatening natural hazard in tropical regions, including the Atlantic Basin. Stronger tropical cyclones are termed hurricanes, and occasional hurricanes are termed 'major hurricanes' if sustained winds are greater than 110 mph or at least Category 3 in the Saffir-Simpson damage scale.[1] These major hurricanes can cause significant damage and fatalities if they make landfall in the United States (Pielke et al., 2008; Blake and Gibney, 2011). NOAA's National Hurricane Center occasionally retires hurricane names that cause damage of tens of billions of US dollars and very high fatalities ranging from the tens to the thousands. Hurricane forecasting and monitoring started systematically around 1870 for the United States, thus the era prior to this is defined as the pre-modern era. However, descriptions of hurricanes in the pre-modern era for the United States date back to the late 1600s (Chenoweth, 2006). Some of these well-known hurricanes, including those in the pre-modern era, remain very pervasive in the memory of public citizens, but many older major storms have been forgotten through time. Many of these major hurricanes would be considered 'worst-case' scenarios by hurricane scholars, emergency management personnel, the insurance industry and policy makers. The focus of this is paper is to provide a long-term objective research approach to studying the memory of pre-modern (pre-1870) major landfalling Atlantic hurricanes for selected areas of the United States coast, dating back several centuries.

The scientific understanding of the major hurricanes that have hit the United States dating back several centuries is fairly well understood through numerous historical hurricane reconstructions that utilise archival data to estimate past intensity and tracks of hurricanes (e.g., Mock, 2004, 2008; Chenoweth, 2006, 2014), as well as a number of paleotempestology studies (e.g., Donnelly et al., 2015).[2] These studies provide a solid scientific basis for the major hurricanes that have existed in the past but do not address consideration of the public hurricane memory through time.

There has been work adopting a more descriptive approach, focusing on the compilation and general description of pre-1870 United States hurricanes as a whole (e.g. Ludlum, 1963), the hurricane history for numerous states (e.g. Rubillo, 2006; Barnes, 2007, 2013; Fraser, 2009) and individual hurricanes with very prominent societal impacts (e.g. Larson, 2000; Salinger, 2009). However, most of these studies have tended to focus on the immediate societal impacts during and shortly after storm occurrence and on the meteorology of storms. To date, aspects of the historical memory of individual past storms over decades and even centuries have been neglected. Aviles (2013), in her study of the 1938 Long Island Express Hurricane, addresses the significance of understanding the public's hurricane memory of older storms leading up to that event, and she qualitatively addressed which hurricanes were most generally 'popular' and which as such became inscribed into public memory. Other work has made use of archival material to explore the more cultural dimensions of historical hurricane events. Jung et al. (2014), for example, demonstrated various quantitative (including non-parametric) statistical approaches to archival historical data in order to contrast public perceptions of female versus male hurricane names. Pielke et al. (2008), in contrast, utilised quantitative data to "normalize hurricane dollar damage" data per storm from 1900 to 2006 for the United States in order to compare how different hurricanes through time would compare with one another if they occurred at the same time.

Although there has been limited scholarship on pre-modern historical storms that relate specifically to memory through time, Rohland (2015) investigated New Orleans hurricanes during the French and Spanish periods of the eighteenth and nineteenth centuries. This work addressed how approaches to mitigation, adaptation and migration changed through this period, highlighting the importance of public memory of past storms in the way in which people prepared for events. Pfister et al. (2010) provided an approach of studying the memory of storms, focusing on three eighteenth-century case studies of (non-hurricane) storms that struck England and the North Sea (1703), west central Europe (1739) and Portugal (1739) through extensive analyses of archival data. They emphasised a twofold approach to exploring the cultural dimensions of particular storm events: first, understanding the meteorological and socioeconomic impacts of each storm and second, the examination of their position within an affected community's or society's cultural memory. The first aspect can be studied through ship logs, newspapers and diaries (e.g., Wheeler, 2003; Mock et al., 2010). The Great Storm of 1703, for example, might be considered in political memory to be a pervasive, 'everlasting' storm, which in turn helped create its legacy. Pfister et al. (2010) extensively describe how the Great Storm occurred at a critical time in England's political history and in a prominent period of development of modern European thought. Moreover, they consider how subsequent popular literary works described the legacy of the Great Storm, thus effectively helping preserve its public storm memory. The Portugal example was one more engrained in culture to preserve its memory, and the French case was complex and somewhat inconclusive. In all such work, however, approaches to studying the memory of storms have been primarily qualitative and based predominantly on the interpretation of a wide range of historical documentary evidence that did not include a critical mass of primary data.

The present study describes an objective approach to the temporal memory of major hurricanes through time, based on known storms derived from previous long-term historical tropical cyclone reconstructions. The study focused on the memory storms for the following regions within the United

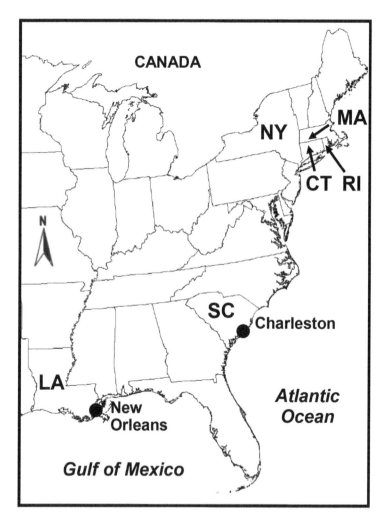

Map 5.1 Map of the states and prominent cities studied for public hurricane memory. Southern New England consists of the states of New York (NY), Connecticut (CT), Rhode Island (RI) and Massachusetts (MA). The other two areas studied are South Carolina (SC) and Louisiana (LA).

States: southern New England located in the northeastern United States; South Carolina located in the southeastern United States along the eastern Atlantic Coast and Louisiana located in the southeastern United States along the Gulf of Mexico (Map 5.1). These regions were selected because of continuous archival data that enabled continuous quantification of what I shall refer to as public hurricane memory. The hurricane memory in this study extends back to 1821 for Southern New England, to 1753 for South Carolina and to 1813 for Louisiana.

Data and methods

Data

Historical data were collected for two different purposes: first for the reconstruction of historical hurricanes and second for the study of hurricane memory. Although this chapter presents the results of the scientific study of major hurricanes that provide the basis for studying the hurricane memory, its focus is on the memory and only highlights are provided here concerning the hurricane reconstruction.[3] Data for this study have been drawn from ship logbooks, newspapers, diaries and various secondary sources. The first three historical data types were used for hurricane reconstruction, while the last three types were used for exploring hurricane memory.

Newspaper material was central to both hurricane reconstruction and the investigation of hurricane memory. Newspapers for hurricane reconstruction came primary from the following state newspaper projects for Massachusetts, Rhode Island, Connecticut, New York, South Carolina, and Louisiana – mostly available on microfilm and in digital format that include those held at the archives of the Library of Congress up through 2015. Several thousand newspaper articles were perused for information on hurricane memory, but many were duplicates of the original article and not used in this study. A total of 321 newspaper articles with over 50 different newspaper titles were utilised in this study. The following examples illustrate references to hurricane memory from newspaper articles, the first being an example for numerous southern New England hurricanes but referring particularly to the Great Gale of 1815 after the occurrence of the Long Island Express hurricane of 1938:

> *Fitchburg Sentinel*, Massachusetts. Oct. 1, 1938, p. 5.
> The tremendous gales of 1723, 1804, 1818, 1821, 1836, 1841, 1851, 1859, 1860, 1869, and some others will be long remembered in certain localities for their severity and the loss of life and property on land and sea which attended them; but neither the memory of man or the annals of our country from the first settlement down to the present time furnish any parallel to the peculiar character of the great gale of September, 1815.

Below is another newspaper example, referring to memory of a hurricane near Charleston, South Carolina, that occurred in 1752:

> *South Carolina Gazette*, June 12, 1770:
>   On Wednesday night last we had a most violent gale of wind, from E.N.E. to S.E. with heavy rains, which has done more damage to the shipping and wharfs, of any that has happened here in the memory of the oldest man living (the hurricane in 1752 only excepted.)

Personal diaries provide invaluable information for hurricane reconstruction and occasionally the longer set of diaries that can cover decades or afford useful insight into hurricane memory. Most of the diaries perused for hurricane memory came from plantations, compiled by farmers. These sources detail day-to-day plantation activities and represent a form of commercial record keeping. Weather is recorded in these sources as it affected the health and productivity of the plantations (Stewart, 1997). Adamson (2015) described how diaries as a source of historical weather information compared to other documents can be more creative, an unbiased account of individuals' interactions with the 'weather world' and, in some cases, can reveal the ways that weather informed the character they presented to the world. For hurricane memory, diaries can provide important information and specific comments on storms not found or biased in newspaper articles. Research for this study utilised diaries held in numerous historical archives and repositories, including New York State Library, Massachusetts Historical Society, Rhode Island Historical Society, New Bedford Public Library, New Bedford Whaling Museum, Nantucket Historical Association, Mystic Seaport Museum, Yale University and the Connecticut State Library. Places for the South include the South Carolina Library, the South Carolina Historical Society and the College of Charleston. Places for Louisiana include the Hill Memorial Library at Louisiana State University, the New Orleans Public Library, New Orleans Notarial Archives and the Historical New Orleans Collection. It is important to note that while several hundred original diaries were perused, only a small number (7 for South Carolina, 1 for New England, and 2 for Louisiana) were found to actually contain references to past storms, and most generally included only brief daily diary entries. However, there is considerable potential to conduct critical mass data searches for New England and Louisiana, particularly the former that has much larger archival databases than South Carolina and Louisiana. The following examples from plantation diaries, located near Charleston, South Carolina, refer to the memory of the hurricane that hit near Charleston, South Carolina in 1752. All three examples refer to the 1752 hurricane.

> Charles Drayton. August 29, 1813-Accounts from Country-Roads utterly impassible. Crops greatly injured, & much destroyed. Waters at D.h. [Drayton House] higher than has been known since 1752. It rose

to the foot of the pigeon house...so that a broad river seem-ed to be between the dwelling & the hill.

William J. Ball (Limerick Plantation): September 26–27, 1822 – Hurricane began at 11 o'clock this night & continued until between 3 to 4 o'clock next morning The greatest hurricane since 1752.

Thomas Porcher Plantation Journal (Pineville, SC). September 28. We had a most violent Storm last Night of wind and Rain I think it must have been as violent here as the Hurricane in 1752. It commenced bet-ween 11 & 12 O clock & continued until about 3 O clock – Fortunately there are no lives lost We have lost a great Number of valuable Pine trees.

Secondary sources – books, research reports and popular websites – have also had an impact on popularising certain hurricanes through time. This study draws on secondary sources that were non-technical and not writ-ten primarily for an academic audience, including popular hurricane books such as *Divine Wind* by Kerry Emanuel (2005) and Dunn and Miller's (1960) *Atlantic Hurricanes*. Prominent state-wide history books, such as those for South Carolina written by Walter Edgar (1998) and Alexander Ramsay (1809) were also consulted as these history books regularly commented on memorable old hurricanes and their impacts.

Hurricane memory

Analyses were conducted separately for each of the three regions: Southern New England, South Carolina and Louisiana. The first step involved tabulating all the major hurricanes. Next, for each major hurricane, any reference to the storm starting the following year up to 2014 was noted, and its simple presence/absence of memory was noted graphically (Figures 5.1, 5.2 and 5.3). Only a simple presence/absence approach to determine general hurricane memory through time was attempted in this study, as opposed to delving into a more detailed context of the storm memory. Due to changes in newspaper frequency, editorships and variable availability of diaries, the author felt that the historical database would not reveal additional reliable details that are at the same level of consistently descriptive reliability for each year through time. Utmost care was taken to avoid circular reasoning of analysing the steps of hurricane reconstruction versus hurricane mem-ory, as both datasets were completely independent of one another. Using the same account for both would lead to biased interpretations (Kates, 1985). A diarist may have reflected back on an earlier storm, but this was not con-sidered circular reasoning.

Once compilation of all the hurricane memory of a major storm was com-pleted and graphed, it was assessed in a simple threefold classification approach (Table 5.1). This simple classification approach of storm's memory through time falls within one of three main categories: persistent memory, intermit-tent memory and lack of memory. Storms with a persistent memory do not

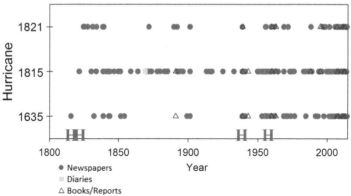

Figure 5.1 Temporal memory of selected major southern New England hurricanes identified from documentary data. "H" refers to the occurrence of a major hurricane in a particular year in order to assess whether it may coincide with increases in hurricane memory.

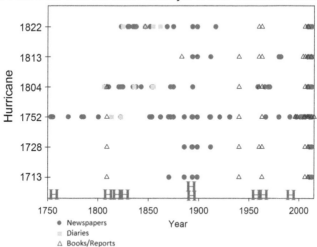

Figure 5.2 Temporal memory of selected major South Carolina hurricanes identified from documentary data. "H" refers to the occurrence of a major hurricane in a particular year in order to assess whether it may coincide with increases in hurricane memory.

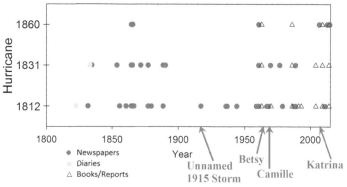

Figure 5.3 Temporal memory of selected major Louisiana hurricanes identified from documentary data. All of these three were selected due to their track passing near New Orleans.

Table 5.1 Three main categories of hurricane memory

1 – Persistent memory:
Hurricane hit major population centres and had massive prominent impact
Relation to some big historical 'legacy' or symbolic images locally
2 – Intermittent memory:
Older hurricane information and interest rediscovered in hurricane anniversaries and recurrences
Different cultural aspects of hurricane impact (e.g. S Carolina nineteenth century rice plantations)
Role of governmental outreach and preparedness (e.g. National Hurricane Center)
Media interest in hurricane/climate debates
3 – Lack of memory (much tougher to interpret):
Hurricane didn't really have much prominent societal impact (perhaps not really a major) or hit remote unpopulated areas
Written records lacking and hurricane quickly 'forgotten', especially for really old hurricanes in remote areas, also few to no deaths

necessarily have reference of memory each year, but they appear not to have been forgotten by the public through time, most probably as a result of having had a significant impact in heavily populated areas. Storms with intermittent memory tended to be those that have been forgotten but then 'rediscovered' numerous times over decadal timeframes, mostly in relation to the popularity of hurricane anniversaries, storm impacts and educational outreach. Storms with a lack of memory simply had very few accounts of the past, and some of these perhaps hit remote areas or may not have really been major hurricanes warranting a full additional scientific reanalysis.

Results

Southern New England

Information on histories of major hurricanes for the period prior to 1870 were drawn mostly from Ludlum (1963), Chenoweth (2006, 2014), and unpublished data from Mock (Marlon JR, Pederson N, Nolan C, Goring S, Shuman B, Booth R, Bartlein PJ, Berke MA, Clifford M, Cook E, Dieffenbacher-Krall A, Hessl A, Bradford Hubeny J, Jackson ST, Marsicek J, McLachlan J, Mock CJ, Moore DJP, Nichols J, Robertson A, Schaefer J, Troet V, Umbanhowar C, Williams JW and Yu Z, Paleoclimate of the northeastern United States during the past 3000 years. *Climate of the Past*, submitted). Overall, major hurricane landfalls are very rare in New England, as simultaneous synoptic weather situations require a strong trough to the west quickly steering a major hurricane tracking from the South along the Atlantic Coast. Only two major hurricanes are clearly evident from documentary records prior to 1870: the September Gale of 1815 that hit near Rhode Island and Massachusetts and the Norfolk-Long Island 1821 hurricane that passed over New York City and went through southern New England. Although previous reanalysis efforts by NOAA currently list a major hurricane as occurring in 1869, Chenoweth (2014) and instrumental data analysed by Mock suggest that it fell quite short of that level, thus it was not included in the major hurricane candidate list for further analysing hurricane memory. The Great Colonial Hurricane of 1635 was also described in some documents and detected in palaeotempestology studies (Donnelly et al., 2015); modelling simulations suggest it was perhaps a Category 4 hurricane upon making landfall in southern New England (Jarvinen, 2006). Therefore, this hurricane, along with those of 1815 and 1821, was further scrutinised in terms of its public hurricane memory.

Results on temporal hurricane memory for southern New England reveal clearly that the 1815 hurricane has been persistent in the public memory, with the other two hurricanes being more intermittent (Figure 5.1). Hurricane memory for this storm was found in 91 newspaper articles, 1 diary entry and 11 secondary accounts. Routinely, New England newspapers reminded the public of this legendary hurricane during hurricane season year on year. A somewhat intermittent and less memorable period in the early 1900s for the 1815 hurricane may be evident, but this is likely attributed to the types of newspapers available in this study and the editorship changes of the newspapers. The 1821 hurricane memory data came from 33 newspaper articles and 9 secondary accounts. This hurricane appeared memorable in the years following its occurrence in the early nineteenth century, related perhaps to its popularisation by William Redfield in his development of the *Law of Storms*, which was widely published and at times mentioned in the newspapers and other publications (e.g., Redfield, 1831). The 1821 hurricane subsequently disappeared from public memory, though it was occasionally mentioned in the later 1800s in relation to the occurrence of subsequent hurricanes. It later returned and proved to be pervasive in hurricane memory starting around the 1950s, largely as a result of the establishment of the National Hurricane Center and its work in building up historical hurricane databases, outreach and hurricane education and the publication of a number of popular hurricane books (e.g. Dunn and Miller, 1960). The popularity of the 1954 New England hurricanes also likely contributed to more pervasive memory. Hurricane memory for the 1635 hurricane was found in 44 newspaper articles and 13 secondary accounts. Memory of the 1635 Great Colonial Hurricane was also prevalent in the early nineteenth century. Temporal changes of its memory follow similar patterns to that of the 1821 hurricane.

South Carolina

The chronologies of past major hurricanes for South Carolina were constructed on the basis of information collected by Mock (2004) and Chenoweth (2006), as well as unpublished data drawn from over 50 ship logbooks, over 50 plantation diaries and letters and over 100 newspaper articles. Six major hurricanes are evident (Figure 5.2). The 1728, 1752 and 1813 hurricanes made landfall near Charleston, South Carolina. The 1713 and 1822 hurricanes made landfall generally about 45 km north of Charleston and just south of Georgetown, and the 1804 hurricane mostly impacted the Beaufort, South Carolina, area near the Georgia border.

The only major hurricane that appeared to possess persistent hurricane memory is the Great Carolina Hurricane of 1752, which also has been described extensively in a number of South Carolina history books (e.g., Ramsay, 1809) and is at times regarded as possibly the strongest hurricane ever to impact the state. Hurricane memory for this storm was found in 37 newspaper articles, 3 diary entries and 2 secondary accounts. Its impacts were felt prominently at Charleston, and it resulted in a storm surge of about 17 feet, often described in recent decades as being comparable with Hurricane Hugo of 1989. Newspapers regularly refer to this event in reflections on past storms during hurricane anniversaries and in general articles on the Atlantic hurricane season. Mercantini (2002) described the pervasive memory of the 1752 hurricane in the political and agricultural impacts leading up to the American Revolution. The memory of the 1804 hurricane (29 newspaper articles, 5 diary entries and 1 secondary account) and that of 1822 (21 newspaper articles, 4 diary entries and 1 secondary account) were quite prevalent in the nineteenth century but mostly intermittent and disappearing during much of the twentieth century, being only more recently remembered after 2000, perhaps in association with a more general growing interest in hurricane history. The nineteenth-century memory of these hurricanes likely relates to cultural aspects of southern plantation agriculture, as these September gales were described as events typical and normal in southern Antebellum culture (Stewart, 1997; Tuten, 2010; Smith, 2012). Only nine newspaper articles and one secondary account were found relating to memory of the 1813 hurricane. This lack of memory for the 1813 hurricane is a bit puzzling, as that storm was a strong major hurricane that hit near Charleston, South Carolina. Perhaps its memory was 'lost' given the context of the War of 1812 the implications of which have gained a more prominent place in popular memory, but no literature or references to date suggest this. Data on hurricane memory for the 1713 and 1728 hurricanes were approximately the same as for the 1813 hurricane. Therefore, these hurricanes did not give rise to a pervasive hurricane memory, the exception being during the 1890s, a period that corresponded to the demise of rice growing and several devastating hurricanes, including two major hurricanes in 1893. The 1893 hurricanes provided the final *coup de grace,* destroying the struggling post-Civil War South Carolina rice agriculture in terms of rice yields and economics (Tuten, 2010), never again enabling rice to be an important agricultural export for South Carolina. The relationship between hurricanes and the demise of the rice culture has become commonplace in general histories of South Carolina (e.g. Rogers, 1970; Edgar, 1998).

Louisiana

Hurricane history information for Louisiana was taken from Mock (2008). Although numerous pre-1870 hurricanes were more evident for Louisiana than for South Carolina and southern New England, it was discovered that finding major hurricane memory information for those events outside of New Orleans was difficult and results of public hurricane memory over time were likely to be incomplete. Several New Orleans newspapers have been published at least daily continuously starting from the early nineteenth century. Therefore, this study focused on three major hurricanes that passed near New Orleans: in 1812, 1831 and 1860. The 1812 hurricane passed closer to New Orleans than any other major hurricane (Mock et al., 2010), while the 1831 hurricane made landfall west of New Orleans towards Grand Island along the central Louisiana coast about 160 km west of New Orleans (Mock, 2008). The 1860 hurricane passed about 80 km east of New Orleans.

   The 1812 hurricane memory was informed by 24 newspaper articles, 1 diary entry and 10 secondary accounts. This is the only Louisiana hurricane that could possibly be classified as having pervasive memory, but one can argue that memory of this event was more intermittent in the early 1800s and early 1900s (Figure 5.3). The more pervasive memory of the 1812 hurricane during the mid-late nineteenth century may be in part due to the occurrence and memory of the Last Island Hurricane of 1856, which, as it contributed to around 200 fatalities and led to wholesale destruction of Last Island, creating two separate islands, was of interest to the public and was thus often mentioned in hurricane recollections (Salinger, 2009). Hurricane memory for the

1831 hurricane was found in 12 newspaper articles, 1 diary and 4 secondary accounts. There is more intermittent memory of this hurricane, like the 1812 hurricane with its memorialisation during the mid-late nineteenth century. However, in the twentieth century, it was largely forgotten, and its memory only re-emerged in the 1950s most probably for the same reasons as for the New England hurricanes described above. The post-1950 increased memory of the 1831 hurricane was also aided by the occurrences of the New Orleans hurricanes of Betsy in 1965, Camille in 1969 and Katrina in 2005. Archival documentation of the 1860 hurricane memory is quite limited – with references in just seven newspaper articles and four secondary accounts. Memory descriptions appeared within recollection lists of numerous past New Orleans hurricanes. Although it was likely to have been a major hurricane with a Louisiana landfall, its impact on New Orleans and populated areas may have been negligible and well below major hurricane status (Dodds et al., 2009).

Conclusions

This study represents a comprehensive objective attempt to investigate the temporal memory of pre-1870 United States major hurricanes for southern New England, South Carolina and Louisiana. Hurricanes preceding 1870 and selected for hurricane memory were based on robust continuous long-term historical tropical cyclone reconstructions. The public hurricane memory of major hurricanes was mostly based on newspaper reports, but includes references drawn from plantation diaries and various other secondary data sources. Results reveal that a threefold classification of storm's memory through time: persistent memory, intermittent memory and lack of memory. Storms with persistent memory such as the 1815 Gale of New England and the Great Carolina Hurricane of 1752 can be associated with prominent societal impacts linked to their storm surges, wind and heavy rainfall within major centres of population. Storms with intermittent memory tend to be forgotten and rediscovered because of the popularity of hurricane anniversaries, culturally mediated storm impacts (such as those described above in South Carolina hurricanes and rice plantations), influential popular books and educational outreach. Storms with a lack of memory are challenging to interpret but tend to be 'forgotten' because the affected areas are less populated and possibly more rural or they took place during periods of major sociopolitical upheaval.

This study can clearly be much expanded with additional archival data, particularly drawn from diaries and newspapers in more rural areas, though sources of information on hurricane memory in such areas are likely to be scarce and challenging to locate – a challenge encountered in the South Carolina case study. However, southern New England has a larger and richer historical database extending back to the early 1700s and thus may provide more unique personal diaries to further study public hurricane memory.

The documentary data on hurricane memory can be further analysed textually, going well beyond the presence/absence of hurricane memory used in this study and utilising detailed content analysis approaches to assess the vocabulary and significance of the hurricane memory. Similar approaches in this study can be applied to other states in the United States as well as to other countries that have hurricane hazards and possess rich continuous archival databases.

This approach is not confined to major hurricanes. Lower category storms can also have societal impacts contributing to many fatalities from inland flooding and poor hurricane mitigation (e.g. the Hong Kong typhoon of 1906, likely about a small Category 1 typhoon with at least 10,000 fatalities). Of particular interest, hurricane/cyclone/typhoon memory of very rare 'Grey Swan' storms (Lin and Emanuel, 2015), which characterise those with return periods of 100 years or longer, could yield very valuable insights into the way in which such events are inscribed into cultural memory. This objective approach to hurricane memory can also be applied to other types of climate extremes. For example, the author has seen many descriptions of the cold air outbreak in the south-eastern USA in February 1835 as recorded in diaries and newspaper accounts up to the present-day, many describing sub-zero Fahrenheit temperatures in the Deep South, which today is considered unheard of and unprecedented. The public memory of this unique cold air outbreak may perhaps be analogous to events such as the 1815 New England hurricane in terms of having pervasive memory through time. Finally, this longer perspective of hurricane memory can define and categorise the legacy of a hurricane. The public generally would know of these famous hurricanes and likely view those as 'worst case scenarios'. Emergency management and other government agencies could find such information useful to convey to the public as an analog or detailed comparison to a concurrent approaching hurricane, and this could be useful for aspects such as risk communication.

Notes

1 A hurricane wind scale, from 1 to 5 rating based on a hurricane's sustained wind speed. See www.nhc.noaa.gov/aboutsshws.php.
2 Paleotempestology is the study of prehistoric storms and tropical cyclones.
3 For detailed methods on how historical major hurricanes were rated, please refer to Mock (2004, 2008) and Chenoweth (2006, 2014).

References

Adamson GCD (2015) "It is very true that where one is at a loss for the subject one talks of the weather": private diaries as information sources in climate research. *WIRES Climate Change*, 6 (6): 599–611. doi:10.1002/wcc.365.

Aviles LB (2013) *Taken by Storm, 1938: A Social and Meteorological History of the Great New England Hurricane.* Boston, MA: American Meteorological Society, 275pp.

Ball WJ (1822) MS Diary at the South Caroliniana Library. Columbia, SC: University of South Carolina.

Barnes J (2007) *Florida's Hurricane History*. Chapel Hill, NC: University of North Carolina Press, 407pp.

Barnes J (2013) *North Carolina's Hurricane History*. Chapel Hill, NC: University of North Carolina Press, 344pp.

Blake ES and Gibney EJ (2011) *The Deadliest, Costliest, and Most Intense United States Tropical Cyclones from 1851 to 2010 (and other frequently requested hurricane facts)*. NOAA Technical Memorandum NWS NHC-6. National Hurricane Center, Miami, FL, 47pp.

Chenoweth M. (2006) A reassessment of historical Atlantic basin tropical cyclone activity, 1700–1855. *Climatic Change*, 7: 169–240.

Chenoweth M (2014) A new compilation of North Atlantic tropical cyclones, 1851–98. *Journal of Climate*, 27 (23): 8674–8685.

Dodds SF, Burnette DJ and Mock CJ (2009) Historical accounts of the drought and hurricane season of 1860. In Dupigny-Giroux LA and Mock CJ (eds.) *Historical Climate Variability and Impacts in North America*. New York, Dordrecht: Springer: 61–77.

Donnelly JP, Hawkes AD, Lane P, MacDonald D, Shuman BN, Toomey MR, van Hengstum PJ and Woodruff DD (2015) Climate forcing of unprecedented intense-hurricane activity in the last 2000 years. *Earth's Future*, 3 (2): 49–65.

Drayton C (1813) *Drayton Papers, 1701–2004*. Charleston, SC: College of Charleston.

Dunn GE and Miller BI (1960) *Atlantic Hurricanes*. Baton Rouge: Louisiana State University Press, 326pp.

Edgar WB (1998) *South Carolina: A History*. Columbia, SC: University of South Carolina Press, 716pp.

Emanuel K (2005) *Divine Wind – The History and Science of Hurricanes*. Oxford: Oxford University Press, 296pp.

Fitchburg Sentinel (1938) Oct. 1, 1938, p. 5. Fitchburg, MA.

Fraser WJ, Jr. (2009) *Lowcountry Hurricanes Three Centuries of Storms at Sea and Ashore*. Athens GA: University of Georgia Press, 368pp.

Jarvinen BR (2006) Storm Tides in Twelve Tropical cyclones (including Four Intense New England Hurricanes). Report, National Hurricane Center, Miami, FL, 99pp.

Jung K, Shavitt S, Viswanathan M and Hilbe JM (2014) Female hurricanes are deadlier than male hurricanes. *Proceedings of the National Academy of Sciences*, 111: 8782–8787.

Kates RW (1985) The interaction of climate and society. In Kates RW, Ausubel JH and Berberian M (eds.) *Climate Impact Assessment*. New York: John Wiley and Sons: 3–37.

Larson E (2000) *Isaac's Storm: A Man, a Time, and the Deadliest Hurricane in History*. New York: Crown Publishers: 323pp.

Lin N and Emanuel K (2015) Grey swan tropical cyclones. *Nature Climate Change*, 6: 106–111.

Ludlum DM (1963) *Early American Hurricanes 1492–1870*. Boston, MA: American Meteorological Society, 198pp.

Marlon et al., Paleoclimate of the north-eastern United States during the past 3000 years. Climate of the Past, submitted.

Mercantini J (2002) The great Carolina hurricane of 1752. *South Carolina Historical Magazine*, 103: 351–365.

Mock CJ (2004) Tropical cyclone reconstructions from documentary records: examples from South Carolina. In Murname RJ and Liu KB (eds.) *Hurricanes and Typhoons: Past, Present and Future*. New York: Columbia University Press: 121–148.

Mock CJ (2008) Tropical cyclone variations in Louisiana, U.S.A., since the late eighteenth century. *Geochemistry Geophysics Geosystems*, 9 (5): doi:10.1029/2007 GC001846.

Mock CJ, Chenoweth M, Altamirano I and Rodgers MD (2010) The great Louisiana hurricane of August 1812. *Bulletin of the American Meteorological Society*, 91: 1653–1663.

Pfister C, Garnier E, Alcoforado MJ, Wheeler D, Luterbacher L, Nunes MF and Taborada JP (2010) The meteorological framework and the cultural memory of three severe winter-storms in early eighteenth-century Europe. *Climatic Change*, 101 (1–2): 281–310.

Pielke RA, Gratz J, Landsea CW, Collins D, Saunders MA and Musulin R (2008) Normalized hurricane damage in the United States: 1900–2005. *Natural Hazards Review*, 9 (1): 29–42.

Ramsay D (1809) *Ramsay's History of South Carolina, from Its First Settlement in 1670 to the Year 1808*. Charleston, SC: David Longworth: 2 vol.

Ravenel TP MS Diary at the South Carolina Historical Society. Charleston, SC.

Redfield WC (1831) Remarks on the prevailing storms of the Atlantic Coast of the North American states. *American Journal of Science and Arts*, 20: 17–51.

Rogers GC (1970) *The History of Georgetown County, South Carolina*. Columbia, SC: University of South Carolina Press, 566pp.

Rohland E (2015) Hurricanes in New Orleans: disaster migration and adaptation, 1718–1794. In Sommer B (ed.) *The Cultural Dynamics of Climate Change in North America*. Leiden/Boston: Brill: 137–158.

Rubillo T (2006) *Hurricane Destruction in South Carolina: Hell and High Water*. Charleston, SC: The History Press, 160pp.

Salinger A (2009) *Island in a Storm: A Rising Sea, a Vanishing Coast, and a Nineteenth-Century Disaster that Warns of a Warmer World*. New York: Perseus Books Group, 290pp.

Smith HR (2012) *Rich swamps and rice grounds: the specialization of inland rice culture in the South Carolina lowcountry, 1670–1861*. PhD dissertation, Department of History, Athens, Georgia: University of Georgia, 319pp.

South Carolina Gazette, June 12, (1770) p. 2. Charleston, South Carolina.

Stewart MA (1997) Let us begin with the weather? Climate, race, and cultural distinctiveness in the American South. In Teich M, Porter R and Gustafsson B (eds) *Nature and Society in Historical Context*. Cambridge: Cambridge University Press: 240–256.

Tuten JH (2010) *Lowcountry Time and Tide: The Fall of the South Carolina Rice Kingdom*. Columbia, SC: University of South Carolina Press, 178pp.

Wheeler D (2003) The Great Storm of November 1703: a new look at the seamen's records. *Weather*, 58 (11): 419–427.

6 "May God place a bridge over the River Tywi"

Interrogating flood perceptions and memories in Welsh medieval poetry

Hywel M. Griffiths, Eurig Salisbury and Stephen Tooth

Introduction

With predicted future increases in the magnitude and frequency of extreme hydrological events such as heavy rainfall, floods and drought (e.g. Fischer and Knutti, 2015), societies will need to adapt to their environmental, socioeconomic, political and cultural impacts (Wilby and Keenan, 2012; Rojas et al., 2013; Arnell et al., 2015). Public perceptions of such impacts and strategies for adaptation and mitigation can be strongly influenced by societal, cultural and historical factors (McEwen and Werritty, 2007; Hulme, 2008; Adger et al., 2016; Devitt and O'Neill, 2016). This can present significant challenges to policy makers and regulatory agencies who traditionally communicate environmental management practices using scientific data alone. Until recently, data from historical cultural sources have not been fully utilised by these actors and are rarely referenced in public discussions of climate change. However, there has been increasing recognition of the relevance and potential of historical ephemera to reconstruct rainfall, floods, drought and related weather phenomena; for example, diaries (Macdonald et al., 2010), newspaper reports (Foulds et al., 2014; Jeffers, 2014), personal correspondence (Endfield and Nash, 2002), estate records (Brázdil et al., 2012; Endfield, 2016), church records (Hall, this volume), tax records (Brázdil et al., 2014), school log-books (Foulds et al., 2014) and creative literature (Macdonald et al., 2010).

The potential value of such sources is twofold. First, they can extend and augment limited instrumental records of rainfall, flooding and drought (Brázdil et al., 2010; Kjeldsen et al., 2014; Benito et al., 2015) that commonly only span decades. By themselves, instrumental records are typically too short and heterogeneous to be used for flood frequency analysis and flood planning and may be misleading if considered in isolation (Wilby and Quinn, 2013). Integrating instrumental records with documentary and other forms of historical evidence, however, can provide a richer and more valuable flood record. In Wales, for instance, historical climatologists are starting to mine Welsh-language historical sources for flood, drought and related climatological data (Macdonald et al., 2010; Jones et al., 2012a; Foulds et al., 2014). Second, such sources can illustrate individual and social perceptions of extreme hydrological events and the environment more generally. Rippon (2009), for example, wrote about how wetlands were perceived as wild and dangerous areas in the medieval period and how their reclamation was seen as a symbol of spiritual improvement, while Galloway (2013) examined coastal communities' reactions to an increased frequency of extreme storms in medieval England.

Although there are numerous examples of floods in the medieval and early modern periods being explained in terms of divine providence (e.g. the Christmas Flood of 1717 on the North Sea Coast; Sundberg, 2015), other societies see floods as much more commonplace. For example, Griffiths and Salisbury (2013) showed that rather than reflecting the divine punishment narrative only, both real and metaphorical floods were often depicted in medieval Welsh poetry as intrusions that can be seen to illuminate the poets' innovative use of genre. In poems of praise and elegies alike, rivers are often used to represent the patron's extended geographical influence, but through the poet's hyperbole in elegies in particular, they are seen to swell up uncontrollably as they are fed by tears of grief. Similar to descriptions of early modern flooding in England analysed by Morgan (2015), these floods were not necessarily described in providential terms. Rather, the ways in which flooding is framed in praise, elegiac and more light-hearted poems suggests that flooding was a rather more common occurrence, certainly for the travelling poets of the nobility. This is echoed in documents relating to the Danube dating from the fifteenth and sixteenth centuries, where the "normality of frequent floods" (Rohr, 2005: 83) is an important feature. Similarly, Hall and Endfield (2016) note that memories and perceptions of historical extreme events, particularly snowy winters, are far from wholly negative; they frequently recall childhood memories with a positive nostalgia.

The complex issue of memory, particularly its relationship with place, has been extensively explored by human geographers and social scientists. Memories can be linked to particular sites, so-called *lieux de mémoire* (Nora, 1989), such as Ground Zero, Hiroshima and Auschwitz, or monuments to people and events such as Lenin and Abraham Lincoln (Hoelscher and Alderman, 2004; Legg, 2007). Memories are frequently political and, in various ways, performative (Alderman, 2010; Johnson, 2012). Studies of memory and poetry in geography (e.g. Griffiths, 2014) have shown that memories are often fragmented and dispersed, rather than focused and singular. Recent work by McEwen and Jones (2012) and McEwen et al. (2014) explored contemporary memories of recent flood events in the Somerset Levels and Gloucestershire, and although Haughton et al. (2015) advise caution regarding the privileging of local above expert knowledge, there is recognition that local and lay flood memories and insights can be very valuable in building community resilience. Krause et al. (2012) have shown that, regardless of the historical period considered, remembering and forgetting floods is an "active and creative process" (p. 128) central to community identity.

Such social memories of hydrological processes, imbued with political meaning and power, can have significant implications for how contemporary societies perceive solutions to potential management issues that climate change is likely to demand. For example, Griffiths (2014) showed that the inherited memories of the politically controversial decision in the 1950s by Liverpool City Corporation to flood the Tryweryn valley, north Wales, in order to provide water for the English city are in some instances linked to contemporary flooding and drought. For instance, during periods of drought, the ruins of the flooded village resurface, reminding people of the injustice of the flooding. Hence, as well as having strong links to place and landscape, memories have a strong temporal element and can vary in relation to hydrological processes. The largely negative memories associated with Tryweryn can be influential in shaping responses to discussions around the best way to manage water at a national scale in contemporary society. Bankoff (2013) has also noted the importance of understanding the historical evolution of water management and engineering for contemporary policy in the context of the coastline of eastern England.

These location-specific cultural and historical contingencies need to be understood in order to enable successful mitigation of, and adaptation to, future climate extremes. As Morgan (2015: 38) puts it, "... culture has a crucial role to play in the construction of floods and their histories". In particular, poetry can be a rich and valuable source of information about weather and flooding. In Welsh-language literature, extreme weather events such as floods and cold summers have been recorded in poems, summer carols and meteorological *englynion* (short strict-metre poems) from the medieval period onwards (Johnston, 2010; Charnell-White, 2011). Farther afield, floods have also been recorded in poetry in Ireland (Collins, 2013) and Australia (Soerjohardjo, 2012). In this chapter we interrogate the historical flood memories recorded in a particular expression of Welsh medieval culture, that of a praise poem – the dominant *genre* in medieval Welsh poetry based on a system of formal patronage in which the poet praised his host or patron – by one of the foremost poets of the fifteenth century, Lewys's Glyn Cothi (*fl.*1447–1489). A new edition and discussion of this poem are presented, and in taking an ecocritical approach (see Rudd, 2007; Johns-Putra, 2016; and in Welsh, Lewis, 2005, 2007, 2008; Evans, 2006; Williams, 2008) and a detailed examination of the context, genre and style of the poem, we investigate the ways in which a single flood event is recorded, described, imagined and remembered. We then evaluate the potential for integrating such artistic expressions with scientific data in order to inform environmental management and community flood resilience.

Lewys Glyn Cothi and the River Tywi floodplain

The 114 km long River Tywi is the longest river flowing entirely within Wales and drains a catchment of 1373 km^2 (see Map 6.1 inset). Rising in the Cambrian Mountains, it flows in a south-westerly direction into Carmarthen Bay. In its middle reaches, the Tywi is a dynamic, meandering gravel-bed river. The floodplain is characterised by suites of river terraces and palaeochannels, indicating that the river has been both vertically and laterally dynamic during the Holocene. Jones et al. (2011a) presented a detailed flood history of the River Tywi, derived from instrumental and documentary records (mainly those listed in the Chronology of British Hydrological Events database – Law et al., 1998) as well as an interpretation of sedimentary evidence. The earliest listed flood occurred in 1767, and although detailed, the documentary record is incomplete (e.g. the 1947 flood is missing). Narrative sources such as poems can be used not only to augment such records, but also to investigate the extent and impacts of flood events and, crucially, how they are inscribed into cultural fabric and memory.

Immediately downstream of the small village and medieval castle of Dryslwyn, the River Dulais joins the trunk stream. Near Felindre, the River Dulais is joined by the small, westward-draining Nant Lais stream (Map 6.1). The poem by Lewys Glyn Cothi discussed below was addressed to patrons in this area *c*.1450–1475.

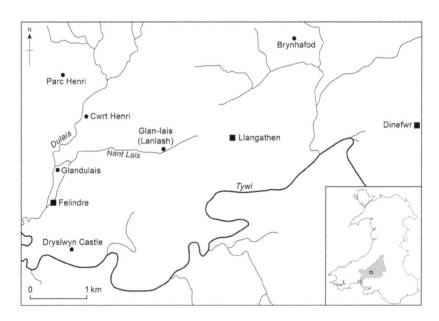

Map 6.1 Map of the middle Tywi valley around Dryslwyn. Inset: the River Tywi catchment.

Text and translation

This poem is found in three manuscript copies, two of which derive from an autograph text that forms part of a large manuscript collection of Lewys's poetry (completed *c*.1475–1480).[1] Figure 6.1 presents a new critical edition based on a detailed transcription of the text and provides for the first time a translation of the poem into English (for the standard edition, see Johnston, 1995: poem 51). It is a praise poem for two of Lewys's most prominent patrons in the Tywi valley, Llywelyn ap Gwilym and Henry ap Gwilym, brothers who lived in the vicinity of Llangathen (Map 6.1) in the second half of the fifteenth century. The poem records a flood on the River Tywi that impeded the poet's journey across the valley from one court to another. Lines 61 and 62 are incomplete in the manuscript.

The poem is in the form of the primary metre of the fifteenth century, the *cywydd deuair hirion*, consisting of seven-syllable lines in rhyming couplets. One of the rhyme-words must be monosyllabic and the other polysyllabic. Every line must be in full *cynghanedd*, a sound-system in Welsh poetry based on intricate patterns of alliteration and internal rhyme. The title in the manuscript source – *kywyd ir.ii.v'der* ('A *cywydd* to the two brothers') – is written along with an ascription in red ink above the poem and refers obliquely to the previous poem, a *cywydd* of praise to Llywelyn ap Gwilym on his own (Johnston, 1995: poem 53). The manuscript also includes one other poem to Llywelyn, a poem to Henry and a poem to their cousin, Owain ap Tomas. Other poems by Lewys to both Llywelyn and Henry, their brother Gwilym Fychan, their father Gwilym, and Henry's son Hywel, are found in other manuscripts (Johnston, 1995: poems 32, 38, 50, 52, 54, 55, 113 and appendices I and II). Both Llywelyn and Henry held offices in local administration during the second half of the fifteenth century, most notably in the castles of Dryslwyn and Dinefwr (Griffiths, 1972). Henry's career in particular seems to have been notable and deserves a detailed study in its own right, but this is beyond the scope of this chapter.

Medieval Welsh poets performed their poems at a feast in front of an audience that usually included the poet's patron, members of the patron's family, other poets and musicians, as well as local people of varying degrees of influence. Musical accompaniment was commonplace, usually on the harp. Very little is known, however, about the specific context of these live performances. When accessing medieval Welsh poems today in both print and online editions,[2] modern readers are usually introduced to the texts by way of an informative title – typically "In praise of [name] of [place name]" – that conveniently places the poem in its geographical and genre-specific context. As would be expected in a largely oral culture, poems recorded in fifteenth-century manuscripts are often untitled, and the few extant titles seem to be either a mark of the poems' renown – flattering monikers attached to the poems at a later date – or (as in Lewys's case) a sign of the scribe's penchant for orderliness. Were the poems verbally introduced and

**Moliant Llywelyn a Henri ap Gwilym
o ddyffryn Tywi gan Lewys Glyn Cothi**

*In praise of Llywelyn and Henry ap Gwilym
of the Tywi valley by Lewys Glyn Cothi*

	O Dduw, ys da o ddeuwr	*By God, two good men*
	Sy'n anheddu deutu'r dŵr:	*dwell on both sides of the water:*
	Llywelyn, erchwyn ein iaith,	*Llywelyn, our nation's protector,*
4	Un rywl yw Henri eilwaith;	*Henry again is of the same standard;*
	Dau froder, un gryfder grym,	*two brothers, one strength of force,*
	Dwy wal i'r wlad o Wilym;	*two of the land's bulwarks descended from Gwilym;*
	Deufur Domas ap Dafydd	*Tomas ap Dafydd's two ramparts*
8	A gaeai'r wlad yn Gaer-ludd.	*who defended the land as if it were London.*
	Ef aeth wyrion Goronwy	*The grandsons of Goronwy*
	O flaen mil filiwn a mwy.	*surpassed a thousand million and more.*
	Ar gân y darogenais	*In a song I foretold*
12	Lenwi y wlad o Lan-lais.	*the filling up of the land from Glan-lais.*
	O dderwen Llan Gathen gynt	*From the oak of Llangathen*
	Imp wedy'r impio ydynt.	*they're a burgeoned offshoot.*
	Llywelyn yw'n impyn ni,	*Llywelyn's our stripling,*
16	A'r unrhyw ŵr yw Henri.	*and Henry's the same.*
	Yfed eu gwin, rhyw fyd gynt,	*They'd cause me, a favourite of yore,*
	Ym mhob awr ym a berynt,	*to drink their wine at every opportunity,*
	A manegi, myn Iago,	*and to narrate, by James,*
20	Ystoriâu Brutus o Dro,	*the tales of Brutus of Troy,*
	A hen gerdd, a henwau gwŷr	*and the earliest poetry, and the names of men*
	A rhieni yr henwyr.	*and of the ancients' ancestors.*
	Ba les? Rhwng deublas y rhain	*To what use? Between their two houses*
24	Mae'r dŵr mwya' o'r dwyrain;	*lies the greatest body of water from the east;*
	Porth Duw boparth i Dywi,	*God's gateway on both sides of the River Tywi,*
	Porthfa yr hindda yw hi.	*she's a haven in fine weather.*
	O'i deupen, fal llanw Menai,	*From both ends, like the flow of the River Menai,*
28	Llenwi yn un llyn a wnâi.	*she filled up like a single lake.*
	Llanw o fôr allan a fydd,	*There will be an outward sea-flow,*
	Llif ennaint oll o fynydd.	*a bath house's entire flow from the mountains.*
	Yr oedd – llyma'r arwyddion –	*These are the signs – she used to have*
32	Arfer hael, ddigrif ar hon,	*a generous, jovial nature,*
	A'i llif a ddyco o'r llall,	*and the flow she bore from one land,*
	Hi a'i dyry i'r wlad arall;	*she transferred it to the next;*
	Hoswi yw hi wedy hyn,	*after this she's a hussy,*
36	A'i hoswïaeth dros ewyn.	*and her thrift is for froth.*
	Ei llif a ddwg ar ei lled	*Her outspread flow carries*
	Ŷd a gwair hyd y gored;	*corn and hay as far as the weir;*
	Oddyno mudo i'r môr	*bearing away from there to the sea*
40	Y derw rhwng ei dau oror;	*the oak trees between her two flanks;*
	Diwreiddio, cwympo y coed	*uprooting, felling the trees*
	A goresgyn y gwrysgoed.	*and overwhelming the thickets.*
	Is y tai dengys Tywi	*Below the houses the River Tywi shows*
44	Maint yw lled ei mantell hi,	*how great is the breadth of her mantle,*
	Mam llifeiriaint berwnaint bas,	*mother to the flood of shallow bubbling streams,*
	Morgymlawdd amryw gamlas,	*sea-surge of many channels,*
	Merch i afon Urddonen,	*daughter of the River Jordan,*
48	Mordwy hwy no llif Noe Hen.	*a greater deluge than Old Noah's flood.*
	Ystréd dros faestir ydyw,	*She's a straight over open country,*
	Ystryd yn fy rhwystro yw.	*she's a highway that impedes me.*
	Tebig, pan fwyf heb ddigiaw,	*On water yonder, when I'm not displeased,*
52	I Dderdri wyf ar ddwr draw.	*I'm like Deirdre.*
	Ef nid âi, rhag ofn y don,	*He wouldn't go, for fear of the wave,*
	I dir nef drwy un afon.	*through any river to heaven's land.*
	Nid af fi Dywi fal dall,	*I won't go into the River Tywi like a blind man,*
56	Neu ddwr gwineuddu arall.	*nor any other dark brown water.*
	Nofio, myn Pedr, nis medrais	*By Peter, I couldn't swim*
	Ym monwes hafn, mwy no Sais.	*in an inlet, no more than an Englishman.*
	Ni charaf, mwy no'r ddafad,	*No more than a sheep, I don't like*
60	Gorwgl byr, gwargul a bad.	*a small, narrow-headed coracle or a boat.*
	Mae 'm bryd …	*It's my desire …*
	Dramwy yno …	*to journey there …*
	O Dduw rhoed einioes i'r ddau;	*May long life be given to them both by God;*
64	Os myn, rhoed oes i minnau,	*if He desires it, may long life be given to me,*
	Ac eisioes yn ein oes ni	*and in our lifetime notwithstanding*
	Y tro Duw bont ar Dywi!	*may God place a bridge over the River Tywi!*

Figure 6.1 Moliant Llywelyn a Henri ap Gwilym o ddyffryn Tywi gan Lewys Glyn Cothi (In Praise of Llywelyn and Henry ap Gwilym of the Tywi valley by Lewys Glyn Cothi).

prefaced before they were performed, and if so, how? Did the poet or some other official of the court announce the poem, perhaps noting the type of poem and its subject matter? Both the manuscript evidence and, in some cases, the poems themselves suggest it is more likely that the poet set about performing his poem without a preamble.

This poem by Lewys Glyn Cothi is a case in point, for the first 22 lines – a full third of the poem – are unremarkable in so far as they conform to the conventional pattern found in most fifteenth-century praise poems. Poems of this kind were usually addressed to individual patrons, but the fact that Lewys is addressing two patrons here is inconsequential, for they were brothers and therefore shared all the main attributes. Both are named, along with their forebears – [G]*wilym*, [T]*omas ap Dafydd, Goronwy* – and both are loosely located in [G]*lan-lais* and *Llan Gathen* in the Tywi valley. They are also praised for surpassing all others as staunch supporters and defenders of the nation, and Lewys states that he values the time spent (perhaps in his youth) in the company of his generous patrons, cultured men who value literature and learning.

The dramatic potential of the poem seems greatly enhanced if it is as-sumed that Lewys's original audience had no prior knowledge of what would come next, for an audience member could easily have been forgiven for thinking that the poem would continue in the same vein – perhaps some 40 or so more lines of general praise centred on Llywelyn's and Henry's continuing patronage. Instead, the praise giving comes to an abrupt halt with *Ba les?* (To what use?), a blunt question that cuts across the beginning of line 23. Metrically, this natural pause could indicate that the caesura in this consonantal *cynghanedd* should fall immediately after [*ll*]*es*, but the first main stress in fact falls on the following word, *rhwng* (between), which em-phasizes this line's awkward rhythm in contrast with the smooth metrical regularity of the opening section. This is a poetic device to reinforce the fact that the River Tywi is in flood, and Lewys therefore cannot travel freely between his patrons' houses. The river has literally come 'between' the two brothers, limiting Lewys's ability to continue to earn a living as a poet in their company. This change to the natural order in the real world is mirrored in the poem, for Lewys does not get back to the business of praising the two brothers until the closing lines, by which point the flood has taken up almost two-thirds of his poem and interrupted its objective, namely to praise.

Nevertheless, the interference of the river in the poem as an obstacle for the eager poet is in itself a form of flattery. That the poem would be successful in its objective was never in doubt, but what sets it apart from Lewys's other poems to Llywelyn and Henry is the fact that here he is play-fully manipulating his audience's expectations. It would have been plainly obvious to anyone listening to Lewys's poem that the two dwellings located in the first two lines of the poem "on both sides of the water" were, at the time of composition, separated by a flood on the River Tywi. However, after his initial and alluringly subtle reference to this body of water, Lewys settles into his usual routine of giving praise in terms of genealogy, region of influence and heroic attributes. Those present may have expected him at first to address the issue at hand, before suspecting he had chosen to ignore it, only to realize by line 23 that he was in fact leading them on for dramatic effect.

The exact location of the two dwellings on opposite sides of the river is open to discussion. Glan-lais and Llangathen, the only two local places named in the poem, were associated with Llywelyn's and Henry's ancestors. The former was a court (today a modern dwelling known as Lan-lash) that is associated in another of Lewys's poems with the brothers' paternal grandfather, Tomas Fychan, and the latter is the church of Llangathen, immediately east of Glan-lais (Map 6.1), near which at least one other court associated with the family was located. Both places are north of the River Tywi, and so too is every court named by Lewys in other poems to both Llywelyn and Henry and to other members of their immediate family. In poems to Llywelyn, Lewys more than once states that his patron lived at Brynhafod, north east of Llangathen; therefore, it is unlikely that Llywelyn lived at Glan-lais when the poem addressed to him and his brother was composed. However, as the court on the south side of the river is not named in the poem, it is unclear whether Glan-lais is the court referred to on the north side or simply the family's ancestral home and therefore worthy of mention as a location common to both patrons. Nevertheless, the court on the north side of the river may have been either Glan-lais or Brynhafod – or possibly Glandulais or Cwrt Henri, two other courts associated with the family – and the other court was probably located somewhere immediately south of the river, adjacent to the family's main region of influence between Dryslwyn and Llangathen.

Lewys's portrayal of the River Tywi in flood could be taken as further evidence of his preoccupation with praise, for it is presented as the antithesis of everything that is praiseworthy in his two patrons. It also frames Lewys's decision to praise Llywelyn and Henry in conventional terms at the beginning of the poem, hereto examined in terms of genre manipulation, in a different light. In lines 3 to 8, both brothers are seen as steadfast protectors of the Welsh nation, *dwy wal* (two bulwarks) and *deufur* (two ramparts) who "defend the land" – typical hyperbolic descriptions of masculinity in medieval praise poems. However, in light of the later passage that describes the destruction wrought by the river, these conventional descriptions seem particularly prescient. The brothers can be seen retrospectively as unmovable 'barriers' or 'earthworks', manmade defences that resist the force of the river, which is made all the more meaningful by the fact that there is no suggestion in the poem that their dwellings had in any way been damaged by the flood.

Nevertheless, there may be more than a hint of danger for Lewys's patrons in his reference to the river's destruction of the natural world. As well as *ŷd* (corn) and *gwair* (hay), the flooded river also carries *derw* (oak trees) to the sea by "uprooting, felling the trees and overwhelming the thickets". In the opening section of praise, Lewys refers in a conventional manner to the two brothers as an *imp* (offshoot) from the "oak of Llangathen", namely their eminent bloodline in that locality (lines 11 to 16). Furthermore, the verb *llenwi* (to fill up) is used both in line 12 to describe the proliferation of the brothers' kin in the Tywi valley and in line 28 to describe the river swelling in the valley as if it were a giant lake. The well-known and occasionally over-worked topos of likening a nobleman's lineage to an oak is therefore used to sobering effect here, for the implication is that even young men in their prime should be wary of the river's indiscriminate power.

In Welsh, *afon* (river) is a feminine noun, and it is therefore not surprising that Lewys refers to the River Tywi as a feminine entity from line 26 onwards (accordingly, the feminine pronouns have been kept in the English translation). Lewys makes use of *dyfalu*, a technique used in Welsh medieval poetry to describe an object, often a gift given to the poet or to his patron, by means of numerous ingenious comparisons. The fact that Lewys uses this technique outside its usual setting, namely poems of request or thanks, is in itself illustrative of Lewys's playful attitude towards genre in this poem. Many of these descriptions are seemingly neutral or even complimentary – the river is a *mam* (mother) to smaller streams and a *merch* (daughter) to the great River Jordan – but gender plays a less favourable role in lines 31 to 36, where the river's fickleness seems to be characterised as a womanly trait. It used to be *hael* (generous) and *[d]igrif* (jovial) – commendable qualities often attributed to noblewomen in the Middle Ages – as it transferred its water in an orderly fashion from one land to the next, much like a charitable patroness would pass her money on to her poet. But in causing a flood and bringing the region to a stand-still, the river is now a *hoswi* (hussy),[3] an immobile mistress of the house-hold who keeps a prudent eye on every penny. This portrayal of the river's changeability in feminine terms – a changeability bordering on the men-strual in terms such as *llif ... ar ei lled* (outward flow) and *rhwng ei dau oror* (between her two flanks) – may be due solely to the gender of the noun in the Welsh language, but it seems more deliberate when viewed in opposition to the masculine ideals outlined for the two brothers at the beginning of the poem. It also serves as a polite reminder to Llywelyn and Henry not to neglect their duties of patronage.

In conclusion, the intrusion of the River Tywi into the orderly lives of fifteenth-century high society is mirrored in Lewys Glyn Cothi's manipulation of genre, for the flood intrudes also in a literary sense on the genre of praise poetry, Lewys's bread and butter as a poet. This complex interplay reinforces Lewys's portrayal of his patrons as learned men, for it is likely that both Llywelyn and Henry, as well as the poem's wider audience, were well attuned to the subtleties of Lewys's remarkable poem. This particular poem is also illustrative of a wider relationship between genre and fluvial landscapes in medieval Welsh poetry (Griffiths and Salisbury, 2013).

Flood perceptions and memory

As the discussion of this new edition of the poem suggests, the poem generally presents the flood negatively. First and foremost, the flood is portrayed as having impeded an important cultural event: the poet claims that he cannot fulfil his duties as a poet, namely travelling from one house to another for patronage. *Ystryd yn fy rhwystro yw* (She's a highway that impedes me), as Lewys says in line 50. As mentioned above, the river before the flood is described as having an *arfer hael, ddigrif* (generous, jovial nature), moving its flow from one area to another. During the flood the river is seen as a still, single lake. This is similar to contemporary attitudes to flooding, namely that water should be encouraged, by engineering and channelisation, to move through areas of high flood risk (especially urban areas) as quickly as possible. The stillness of the river mirrors the stillness of the poet, stranded on one side of the river. The impacts on crops and trees are also described, and as addressed above there is a note of caution for the two brothers in this description. During the flood, the river becomes a source of fear for the poet: he compares himself to Deirdre, the Irish legendary figure, seen here as a male hero who was afraid of water. The flood water is *gwineuddu* (dark brown), the poet cannot swim, and he does not like the thought of using a coracle or a boat. Nevertheless, although the flood is likely to have restricted Lewys's short-term soliciting of patronage in the Tywi valley, the apparent seriousness of the flood does not, ironically, prevent Lewys from fulfilling at least some of his duties as a poet. Consequently, the poem presents a complex perception of flooding. The flood is superficially described as negative, yet its very existence is creatively manipulated to form a basis for imparting praise. This is in common with flood perceptions in other Welsh medieval poems such as those by Dafydd ap Gwilym and Guto'r Glyn (Griffiths and Salisbury, 2013).

The poem includes four references to religious figures, namely St James, Noah, St Peter and God. Nevertheless, the flood itself is neither seen as providential nor as divine punishment for the sins of mankind. Rather than elicit a providential response based on divine punishment in an attempt to secure its societal memory (*cf.* Kempe, 2007), this flood on the River Tywi inspired a more measured response. This suggests that, although significant enough to be one of the main subjects of the poem, the flood was not an uncommon occurrence in the poet's experience either at this location or elsewhere. Rather than calling on God to spare the community from future flooding and/or calling on his fellow countrymen to mend their sinful ways to avoid incurring God's providential wrath in the future, Lewys simply calls for God to place a bridge over the river, accepting flooding as inevitable. This mirrors the findings from Morgan's (2015: 50) investigations of the perceptions of flooding in England in the early modern period, namely that, although there was an "unwritten structural belief in God", not all descriptions of flooding were couched in terms of divine providence.

It could be argued that the poem is local, if not hyperlocal. The poem is clearly located in a very specific area of a very specific river catchment, and although the poet does make connections with the wider catchment (by referring to the mountains and the sea, for example) and with Wales and the world (by referring to the Menai Strait and the River Jordan), the poem's focus is on the impacts of the flood in a very small location. The poet localises it by naming both a court (Glan-lais) and a nearby settlement (Llangathen) and, crucially, by highlighting the patrons' lineage in that area, by naming the father and other ancestors. Lewys also ends by calling for a local solution to the flood problem – a bridge over the river (although it is likely that proposing such an undertaking is simply another way of underlining how much he wanted to cross the river to be with both his patrons, rather than any wish to see an actual bridge built). As Morgan (2015: 50) puts it, the flood was 'locally negotiated' and provides an example of local weather narrative and knowledge (Veale et al., 2014).

Considered alongside evidence from other medieval poems (Griffiths and Salisbury, 2013) and contemporary Welsh poetry (Griffiths, 2014), Lewys's poem points to a society where water occupies a significant role both as an inspiration for creative activity and as a device and metaphor in stories and poems. As such, it is similar to Schama's (1988) notion of a 'hydrographic culture'. Morgan (2015: 50) describes hydrographic societies as those that "expressed their hydrology and relationships to it in writing", and Kempe (2007) talks of a similar concept, namely the 'amphibian societies'. Morgan (2015) found that early modern societies in England 'encoded' their relationships to water in a number of genres as a way of coping with flood impacts. In this poem, Lewys makes use of a literary topos common in fifteenth-century Welsh poetry, namely *llif Noe* ('Noah's flood – line 48). This combination of words is found in many poems of this period, specifically elegies. For example, in a number of his elegies and poems of compassion for his patrons, Guto'r Glyn (*c*.1415–1490) likened the tears that he shed to Noah's flood, saying that *Llif Noe yw'r llefain a wnawn* (The tears I shed were Noah's flood) (Griffiths and Salisbury, 2013).

The flood on the River Tywi probably occurred sometime between 1450 and 1475, and the mid-fifteenth century experienced periods of negative North Atlantic Oscillation (NAO) index[4] usually associated with flood-rich periods (see below). Although more work is needed to interrogate the available body of literature more thoroughly, it could be argued that the predominant environmental conditions of the period were reflected in its poetic imagery and that medieval (and contemporary) Welsh society was hydrographic.

Compared to other parts of the UK and Europe, there is a notable dearth of epigraphic markers (flood levels inscribed on structures – Macdonald, 2007) on Welsh bridges, churches and other buildings. However, floods in Wales are commonly remembered in other ways. The memory of this single mid-fifteenth century, probably relatively common flood, for instance, has been encoded in this single poem. Its subsequent transmission across the generations is a result of a complex series of occurrences. First, the poet has drawn inspiration from the flood itself and has creatively represented it in a complex poem, musing on the flood's meaning for himself, his patron and his audience, in effect playing with the poetic possibilities afforded by it in the creative process. Next, the poet performed the poem in front of an audience, possibly more than once, thus transmitting and reinforcing the memory of the flood in their minds. Then, the poet himself recorded it in a manuscript that became known to a number of influential poets and humanist scholars during the following centuries – including the notable patron Gruffudd Dwnn (*c*.1500–1570), the great poet and herald Gruffudd Hiraethog (d. 1564) and three of the most learned transcribers and manuscript collectors of the early modern period, Thomas Wiliems of Trefriw (*c*.1545/6–*c*.1623), John Davies of Mallwyd (*c*.1570–1644) and Robert Vaughan of Hengwrt (1591/2–1667) – before it eventually became part of the vast collection of manuscripts at the National Library of Wales in Aberystwyth. It is not known to what extent the flood or the poem was remembered locally after the fifteenth century as part of an oral or a scribal tradition. It is likely, then, that the memory of the flood waxed and waned across the centuries, recalled and recycled at particular times depending on access and interest.

Welsh medieval poetry as a source of palaeohydrological and palaeoenvironmental information

Considerable potential exists for integrating creative flood memories represented in Welsh medieval poems with geomorphological and sedimentological evidence to better understand flood histories and landscape development. This potential could be explored in the Tywi valley and beyond. Jones et al. (2011a, b) undertook detailed geomorphological mapping in the middle Tywi valley, including at the confluence of the River Dulais tributary and constructed a detailed flood history using a combination of instrumental data, data from the Chronology of British Hydrological Events database, and proxies derived from the analysis of geochemical element ratios in floodplain cores. These data show a number of flood-rich periods related to negative phases of the NAO index, and Jones et al. (2011a) even report an example of a flood that apparently dates from the late fifteenth century, although the potential uncertainty in the method needs to be considered when attempting to relate geochemical evidence to a specific flood.

More broadly, Macklin et al. (2014) have shown that the early medieval period was characterised by rapid and significant anthropogenic changes to Welsh floodplain environments. As a result of the adoption of agriculture and increased soil erosion, the medieval period saw accelerated rates of floodplain sedimentation, with fine sediments filling river channels and changing the previously multi-channel, probably densely wooded, river systems into single channel, less vegetated meandering systems with remnants of former channels across the floodplain. In some areas, this was compounded by early channelization efforts. The combined effects were to constrain the position of many river channels in Wales and the UK more generally, but former channels would be reactivated during floods. This is perhaps best illustrated in this poem by reference to the *Morgymlawdd amryw gamlas* (sea-surge of many channels), potentially suggesting the existence of numerous 'shallow' (*bas*) channels on a wooded valley floor (e.g. similar to the parts of the Gearagh River in Ireland – Harwood and Brown, 1993).

In Lewys's poem, the physical impacts of the flood are vividly described. This is particularly true in terms of the impact on agricultural crops and natural riparian vegetation. For example *Ei llif a ddwg ar ei lled / Ŷd a gwair hyd y gored* (Her outspread flow carries/corn and hay as far as the weir) is a clear description of the flood's impact, and the trees are said by the poet to have been 'uprooted' (*diwreiddio*) and felled, and the oaks carried to the sea – *Oddyno mudo i'r môr / Y derw rhwng ei dau oror* (bearing away from there to the sea/the oak trees between here two flanks). This suggests that the River Tywi included significant amounts of large woody debris during the medieval period, which stands in contrast to the limited amounts in the contemporary river. Overall, the poem records and transmits not only a memory of flooding, but also the geomorphology of the medieval floodplain including observations of climatically and anthropogenically driven changes.

Potential for geographical outreach and education

In 2011, the National Strategy for Flood Risk and Coastal Erosion Risk Management was published (Welsh Assembly Government, 2011). One of its overarching objectives was 'raising awareness of and engaging people in the response to flood and coastal erosion risk' (p. 21). Subsequent significant flood events in the UK, such as those in north Ceredigion (June 2012), the Somerset Levels (winter of 2013–2014) and the north of England and Scotland (winter of 2015–2016) have brought the need for community flood management and flood resilience into even sharper focus. However, responses from media and government following these sorts of events frequently talk about their 'unprecedented' nature, suggesting a lack of understanding in public discourse around the nature of flood histories and flood risk. Developing an improved understanding and public engagement with climate change and its impacts (Lorenzoni et al., 2007) could be highly valuable for enhancing community resilience and preparedness to future hydrological extremes.

Hydrological extremes such as floods are partly meteorological phenomena, but as Lewys's poem illustrates, they are also linked with geomorphological changes, namely changes to channels and floodplains as well as associated riparian vegetation and potentially infrastructure. Understanding geomorphology is thus key to understanding the changing nature of flood risk as a result of climate change, but Tooth (2009) and Woodward (2015) have expressed concern regarding the diminishing visibility of geomorphology as a term in the public consciousness. As a potential solution, Tooth et al. (2016) argue for a greater engagement between geomorphologists and artists, including poets, to communicate data and concepts. Jones et al. (2012b) have also explored the value of the arts in engaging with vulnerable 'watery' landscapes and communities, while McEwen and Jones (2012) and McEwen et al. (2014) have noted that the integration of the arts and humanities with the natural and social sciences is key for the development of integrated, community-based and sustainable resilience. Against this backdrop, medieval literature such as Lewys's poem, with its vivid, engaging descriptions of a significant flood event and its geomorphological effects could be a powerful tool for geographical education and outreach. Indeed, recent discussions around flood heritage (e.g. see 'Floods as Heritage' conference, Limoges, 2015)[5] have explored the potential for promoting historical floods as sites of heritage to be experienced by the public.

In Wales, strict-metre poetry continues to be practised by poets of all ages and is performed at readings and festivals (most notably the National Eisteddfod of Wales) and on radio and television. Middle Welsh can be relatively easily understood by speakers of Modern Welsh and/or easily translated into Modern Welsh (and therefore into other languages). Presenting such an historical account of a damaging flood alongside informative and appropriate scientific data or materials (e.g. flood series, geomorphological maps, images from floodplain and sediment cores), perhaps through an information board, leaflet, website or walking tour podcast has great potential to engage and educate the public. Changing individual and community perceptions is part of developing societal resilience to extreme events (Adger et al., 2005). If presented in a spatially and culturally relevant way, this specific poem, and other environmentally focused poems from the medieval period (as well as the early modern and modern periods), could help change the perceptions of those living and working on floodplains in Wales regarding the nature of flood risk.

Conclusions

Our analysis of Lewys Glyn Cothi's poem has shown how such historical records can augment, extend and enrich instrumental sources, other documentary sources and geomorphological evidence. They can provide an alternative perspective on the climate of the past as well as a human voice to complement scientific data. As a result, these historical literary sources could potentially be used to improve public understanding of flood risk. Although Lewys's plea for a solution to his flood problem – namely a bridge over the River Tywi – is unlikely to have been literal, it is conceivable that his poem could contribute, over five centuries later, to the development of resilient and sustainable solutions to contemporary flood problems. Although this poem illustrates the fragmented, dispersed and serendipitous nature of the recording and encoding of flood memories (due to a "complex combination of various mechanisms" – Hall, this volume), such sources can potentially be integrated with the high-resolution flood chronologies provided by sedimentological and geochemical investigations. Griffiths and Salisbury (2013) have begun the process, but much more needs to be done to explore these seemingly disparate sources. In summary, therefore, this paper underlines the unrealised potential of medieval Welsh literary sources, both as palaeoenvironmental data and for the insights they offer into human-environment interactions and societal perceptions of channels, floodplains and floods. It also suggests that, in common with other historical and contemporary societies in the UK and Europe, medieval Welsh societies were 'hydrographic' in the sense that water had the potential to permeate their cultural expressions, especially by way of poets' innovative use of genre. Future work should focus on systematically reviewing all available Welsh medieval and later poetry in order to identify the varied and changing ways in which rivers, floods and drought were incorporated into the local and national culture.

Acknowledgements

The authors are grateful to Antony Smith for assistance with producing the figures.

Notes

1 National Library of Wales, Peniarth MS 109: 172–174.
2 See Dafydd ap Gwilym project website, www.dafyddapgwilym.net (accessed 30 June 2016); Guto'r Glyn project website, www.gutorglyn.net (accessed 30 June 2016).
3 *Hoswi* probably derives from the Middle English *hussy*, see *Oxford English Dictionary Online* s.v. *hussy* 1 (obsolete) 'The mistress of a household; a thrifty woman'.
4 The North Atlantic Oscillation index is based on the difference between surface pressure in the Azores and Iceland. The NAO index is described as being positive if the Azores High is strong and the Icelandic Low is deep and negative when the opposite pattern is observed. In the UK, a positive NAO index is associated with higher than normal winter precipitation, but some of the largest floods of the past 60 years have been associated with a negative NAO index, caused by convective summer storms or slow-moving fronts (Macklin and Rumsby, 2007).
5 Floods as Heritage Conference Website, www.unilim.fr/geolab/2015/09/14/atelier-la-crue-linondation-un-patrimoine/ (accessed 30 June, 2016).

References

Adger WN, Arnell NW and Tompkins EL (2005) Successful adaptation to climate change across scales. *Global Environmental Change*, 15 (2): 77–86.

Adger WN, Quinn T, Lorenzoni I and Murphy C (2016) Sharing the pain: perceptions of fairness affect private and public response to hazards. *Annals of the American Association of Geographers*, 106 (5): 1079–1096.

Alderman DH (2010) Surrogation and the politics of remembering slavery in Savannah, Georgia. *Journal of Historical Geography*, 36 (1): 90–101.

Arnell NW, Halliday SJ, Battarbee RW, Skeffington RA, and Wade AJ (2015) The implications of climate change for the water environment in England. *Progress in Physical Geography*, 39 (1): 93–120.

Bankoff G (2013) The 'English Lowlands' and the North Sea basin system: a history of shared risk. *Environment and History*, 19 (1): 3–37.

Benito G, Brázdil R, Herget J, and Machado MJ (2015) Quantitative historical hydrology in Europe. *Hydrology and Earth System Science*, 19: 3517–3539.

Brázdil R, Chromá K, Řezníčková L, Valášek H, Dolák L, Stachoň Z, Soukalová E and Dobrovolný P (2014) The use of taxation records in assessing historical floods in South Moravia, Czech Republic. *Hydrology and Earth System Sciences*, 18 (10): 3873–3889.

Brázdil R, Chromá K, Valášek H and Dolák L (2012) Hydrometeorological extremes derived from taxation records for south-eastern Moravia, Czech Republic, 1751–1900 AD. *Climate of the Past*, 8 (2): 467–481.

Brázdil R, Dobrovolný P, Luterbacher J, Moberg A, Pfister C, Wheeler D and Zorita E (2010) European climate of the past 500 years: new challenges for historical climatology. *Climatic Change*, 101 (1–2): 7–40.

Charnell-White C (2011) Cofio'r tywydd yng Nghymru: Casgliad Thomas Evans, Hendreforfudd o englynion meteorolegol. *Llên Cymru*, 34: 62–87.

Collins L (2013) The frosty winters of Ireland: poems of climate crisis 1739–41. *Journal of Ecocriticism*, 5 (2): 1–11.

Devitt C and O'Neill E (2016) The framing of two major flood episodes in the Irish print news media: implications for societal adaptation to living with flood risk. *Public Understanding of Science*. doi:10.1177/0963662516636041.

Endfield GH (2016) Historical narratives of weather extremes in the UK. *Geography*, 101 (2): 93–99.

Endfield GH and Nash DJ (2002) Drought, desiccation and discourse: missionary correspondence and nineteenth-century climate change in central southern Africa. *The Geographical Journal*, 168 (1): 33–47.

Evans DF (2006) 'Cyngor y Bioden': ecoleg a llenyddiaeth Gymraeg. *Llenyddiaeth mewn Theori*, 1: 41–79.

Fischer EM and Knutti R (2015) Anthropogenic contribution to global occurrence of heavy-precipitation and high-temperature extremes. *Nature Climate Change*, 5 (6): 560–564.

Foulds SA, Griffiths HM, Macklin MG and Brewer PA (2014) Geomorphological records of extreme floods and their relationship to decadal-scale climate change. *Geomorphology*, 216: 193–207.

Galloway JA (2013) Coastal flooding and socioeconomic change in eastern England in the later middle ages. *Environment and History*, 19 (2): 173–207.

Griffiths RA (1972) *The Principality of Wales in the Later Middle Ages: The Structure and Personnel of Government*. Cardiff: University of Wales Press, 634pp.

Griffiths HM (2014) Water under the bridge? Nature, memory and hydropolitics. *Cultural Geographies*, 21 (3): 449–474.

Griffiths HM and Salisbury TE (2013) 'The tears I shed were Noah's flood': medieval genre, floods and the fluvial landscape in the poetry of Guto'r Glyn. *Journal of Historical Geography*, 40: 94–104.

Hall A (2017) Remembering in God's name: the role of the church and community institutions in the aftermath and commemoration of floods. In Endfield GH and Veale L (eds.) *Cultural Histories, Memories and Extreme Weather: A Historical Geography Perspective.* New York: Routledge: pp. 112–132.

Hall A and Endfield G (2016) 'Snow scenes': exploring the role of memory and place in commemorating extreme winters. *Weather, Climate, and Society,* 8 (1): 5–19.

Harwood K and Brown AG (1993) Fluvial processes in a forested anastomosing river: flood partitioning and changing flow patterns. *Earth Surface Processes and Landforms,* 18 (8): 741–748.

Haughton G, Bankoff G, and Coulthard T (2015) In search of 'lost' knowledge and outsourced expertise in flood risk management. *Transactions of the Institute of British Geographers,* 40 (3): 375–386.

Hoelscher S and Alderman DH (2004) Memory and place: geographies of a critical relationship. *Social and Cultural Geography,* 5 (3): 347–355.

Hulme M (2008) Geographical work at the boundaries of climate change. *Transactions of the Institute of British Geographers,* 33 (1): 5–11.

Jeffers JM (2014) Environmental knowledge and human experience: using a historical analysis of flooding in Ireland to challenge contemporary risk narratives and develop creative policy alternatives. *Environmental Hazards,* 13 (3): 229–247.

Johns-Putra A (2016) Climate change in literature and literary studies: from cli-fi, climate change theater and ecopoetry to ecocriticism and climate change criticism. *Wiley Interdisciplinary Reviews: Climate Change,* 7 (2): 266–282.

Johnson NC (2012) The contours of memory in post-conflict societies: enacting public remembrance of the bomb in Omagh, Northern Ireland. *Cultural Geographies,* 19 (2): 237–258.

Johnston D (1995) *Gwaith Lewys Glyn Cothi.* Cardiff: University of Wales Press, 684pp.

Johnston D (2010) Tywydd eithafol a thrychineb naturiol mewn dwy farwnad gan Iolo Goch. *Llên Cymru,* 33: 51–60.

Jones AF, Brewer PA and Macklin MG (2011a) *Reconstructing Late Holocene flood records in the Tywi catchment.* Final report for the Dyfed Archaeological Trust Exploration Tywi! Project, 55pp.

Jones AF, Brewer PA, Macklin MG, Swain CH, Bird G, Griffiths HM and Yorke L (2011b) *Late Quaternary River Development and Archaeology of the Middle Tywi Valley.* Final report for the Dyfed Archaeological Trust Exploration Tywi! Project, 97pp.

Jones CA, Davies SJ and Macdonald N (2012a) Examining the social consequences of extreme weather: the outcomes of the 1946/1947 winter in upland Wales, UK. *Climatic Change,* 113 (1): 35–53.

Jones O, Read S and Wylie J (2012b) Unsettled and unsettling landscapes: exchanges by Jones, Read and Wylie about living with rivers and flooding, watery landscapes in an era of climate change. *Journal of Arts & Communities,* 4 (1–2): 76–99.

Kempe M (2007) 'Mind the next flood!' Memories of natural disasters in Northern Germany from the sixteenth century to the present. *The Medieval History Journal,* 10 (1–2): 327–354.

Kjeldsen TR, Macdonald N, Lang M, Mediero L, Albuquerque T, Bogdanowicz E, Brázdil R, Castellarin A, David V, Fleig A and Gül GO (2014) Documentary evidence of past floods in Europe and their utility in flood frequency estimation. *Journal of Hydrology,* 517: 963–973.

Krause F, Garde-Hansen J and Whyte N (2012) Flood memories – media, narratives and remembrance of wet landscapes in England. *Journal of Arts & Communities*, 4 (1–2): 128–142.

Law FM, Black AR, Scarrott RMJ, Miller JB and Bayliss AC (1998) British Chronology of Hydrological Events. Available from: www.dundee.ac.uk/geography/cbhe/ (accessed 19 August 2006).

Legg S (2007) Reviewing geographies of memory/forgetting. *Environment and Planning A*, 39 (2): 456–466.

Lewis BJ (2005) 'Llafurfaith a waith a weinyddaf': dulliau'r gogynfeirdd o agor cerdd. *Llên Cymru*, 28: 1–25.

Lewis BJ (2007) Genre a dieithrwch yn y cynfeirdd: achos 'Claf Abercuawg'. *Llenyddiaeth mewn Theori*, 2: 1–35.

Lewis BJ (2008) Bardd natur yn darllen bardd y ddinas? Dafydd ap Gwilym, 'Y Don ar Afon Dyfi', ac Ofydd, *Amores*, III.6. *Llên Cymru*, 31: 1–22.

Lorenzoni I, Nicholson-Cole S and Whitmarsh L (2007) Barriers perceived to engaging with climate change among the UK public and their policy implications. *Global Environmental Change*, 17 (3): 445–459.

Macdonald N (2007) Epigraphic records: a valuable resource in re-assessing flood risk and longterm climate variability. *Environmental History*, 12: 136–140

Macdonald N, Jones CA, Davies SJ and Charnell-White C (2010) Historical weather accounts from Wales: an assessment of their potential for reconstructing climate. *Weather*, 65 (3): 72–81.

Macklin MG and Rumsby BT (2007) Changing climate and extreme floods in the British uplands. *Transactions of the Institute of British Geographers*, 32 (2): 168–186.

Macklin MG, Lewin J and Jones AF (2014) Anthropogenic alluvium: an evidence-based meta-analysis for the UK Holocene. *Anthropocene*, 6: 26–38.

McEwen L and Jones O (2012) Building local/lay flood knowledges into community flood resilience planning after the July 2007 floods, Gloucestershire, UK. *Hydrology Research*, 43 (5): 675–688.

McEwen L, Jones O and Robertson I (2014) 'A glorious time?' Some reflections on flooding in the Somerset Levels. *The Geographical Journal*, 180 (4): 326–337.

McEwen J and Werritty A (2007) 'The Muckle Spate of 1829': the physical and societal impact of a catastrophic flood on the River Findhorn, Scottish Highlands. *Transactions of the Institute of British Geographers*, 32 (1): 66–89.

Morgan JE (2015) Understanding flooding in early modern England. *Journal of Historical Geography*, 50: 37–50.

Nora P (1989) Between memory and history: Les lieux de mémoire. *Representations*, 26: 7–24.

Rippon S (2009) 'Uncommonly rich and fertile' or 'not very salubrious'? The perception and value of wetland landscapes. *Landscapes*, 10 (1): 39–60.

Rohr, C (2005) The Danube floods and their human response and perception (14th to 17th C). *History of Meteorology*, 2: 71–86.

Rojas R, Feyen L and Watkiss P (2013) Climate change and river floods in the European Union: socio-economic consequences and the costs and benefits of adaptation. *Global Environmental Change*, 23: 1737–1751.

Rudd G (2007) *Greenery: Ecocritical Readings of Late Medieval English Literature*. Manchester: Manchester University Press, 221pp.

Schama S (1988) *The Embarrassment of Riches: An Interpretation of Dutch Culture in the Golden Age.* London: Collins, 698pp.

Soerjohardjo W (2012) Remembering Yarrie: an indigenous Australian and the 1852 Gundagai flood. *Public History Review*, 19: 120–129.

Sundberg A (2015) Claiming the past: history, memory, and innovation following the Christmas flood of 1717. *Environmental History*, 20 (2): 238–261.

Tooth S (2009) Invisible geomorphology? *Earth Surface Processes and Landforms*, 34 (5): 752–754.

Tooth S, Viles H, Dickinson S, Dixon SJ, Falcini A, Griffiths HM, Hawkins H, Lloyd Jones J, Ruddock J, Thorndycraft V and Whalley B (2016) Visualizing geomorphology: improving communication of data and concepts through engagement with the arts. *Earth Surface Processes and Landforms*, 41 (12): 1793–1796. doi:10.1002/esp.3990.

Veale L, Endfield G and Naylor S (2014) Knowing weather in place: the Helm Wind of Cross Fell. *Journal of Historical Geography*, 45: 25–37.

Welsh Assembly Government (2011) National Strategy for Flood Risk and Coastal Erosion Risk Management, Welsh Assembly Government, 85pp.

Wilby RL and Keenan R (2012) Adapting to flood risk under climate change. *Progress in Physical Geography*, 36 (3): 348–378.

Wilby RL and Quinn NW (2013) Reconstructing multi-decadal variations in fluvial flood risk using atmospheric circulation patterns. *Journal of Hydrology*, 487: 109–121.

Williams H (2008) Ecofeirniadaeth i'r Celtiaid. *Llenyddiaeth mewn Theori*, 3: 1–28.

Woodward J (2015) Is geomorphology sleepwalking into oblivion? *Earth Surface Processes and Landforms*, 40 (5): 706–709.

7 Remembering in God's name

The role of the church and community institutions in the aftermath and commemoration of floods

Alexander Hall

Introduction

Many academic studies of past disasters are case-study focused and often do not consider the longer-term and broader social contexts in which the catastrophe may have occurred (e.g. Reilly, 2009).[1] Such an approach often treats extreme weather-caused disasters as non-recurring, one-off catastrophic events, which briefly cause societal breakdown and then disappear.[2]

What happens if we begin to consider these extreme weather events as cyclical occurrences, which have repeatedly challenged specific communities throughout their history? If we take such a viewpoint and consider the longer history of, for example, a coastal community vulnerable to flooding, we begin to see that perhaps the social structures that precede the disaster and seem to influence so greatly a community's ability to cope in its aftermath, have themselves been shaped by earlier flood events that afflicted the same community.

Informed by both the nascent literature on cultural understandings of climate, weather and extreme meteorological events and the widespread literature on disaster risk reduction, this chapter aims to explore how, for one specific historical flood, local churches responded to immediate community needs, increasing social resilience. Further, given that the incorporation of extreme weather events into a community's long-term social memory is important for the community's perception of future risks, adaptive development of infrastructure and ongoing social resilience (Blaikie et al., 2014: 330–346), the chapter also aims to explore what role, if any, these religious institutions played in commemorating and embedding the floods within a longer historical narrative of the region.

The chapter focuses on the town of King's Lynn in Norfolk, south east England (Map 7.1) and the devastating flooding that occurred there as part of the widespread North Sea Floods of 1953. King's Lynn is a seaport situated on the tidal stretch of the river Great Ouse, close to where it enters the estuary and bay known as The Wash (Hunter-Blair, 2001). The town lies on a greatly altered stretch of the river channel, which was originally diverted in the thirteenth century (Richards, 1812: 10–13). The Great Ouse empties much of the former marsh lowland of The Fens, which has been artificially drained by humans since at least the seventeenth century (Knittl, 2007). Due to its specific topography and its location within a water regime long altered by human intervention, King's Lynn has a long history of severe flood events caused by excessive rainfall, overland flow, storm surges and other incursions from the sea (Pollard, 1978: 15; Lamb and Frydendahl, 1991).

The narrative presented in this chapter is drawn from primary research into the records of three churches in King's Lynn (St Margaret's, Stepney and Union Chapel), local and regional newspapers (the *Lynn News & Advertiser* and the *Eastern Daily Press*) and local and national government records from 1953 up to the 1980s. The regional newspapers are held by the British Library in London and are a rich and often-overlooked source of information on weather history. The church records, held at the Norfolk Record Office (NRO) in Norwich, and the government records, held at the National Archives London, whilst containing less content related directly to meteorological conditions, still present an under-studied resource for those scholars prepared to persevere. By using these archival records to build a picture of the role local churches played in the aftermath and commemoration of the 1953 floods, and by putting these records into dialogue with scholarship on community resilience and community memory of disaster, this chapter aims to show how, in one historical and cultural context, religious institutions were crucial for social cohesion, community resilience and the long-term normalisation of flood events in the region.

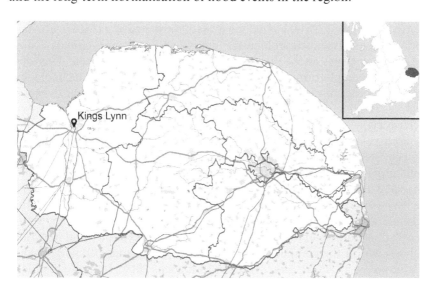

Map 7.1 Map showing the location of King's Lynn in the county of Norfolk and the county's location within England (inset).

Source: Image contains Ordnance Survey data © Crown copyright and database right and is licensed under a Creative Commons Attribution-Share Alike 3.0 Unported license.

In drawing a picture of the role of the churches in King's Lynn during and after the North Sea Floods of 1953, this chapter first introduces the details of the horrific events that began to unfold on 31 January 1953. The chapter then details how the floods affected King's Lynn directly and what role the churches played in the immediate disaster response and recovery of the community. Finally, the chapter explores how the floods were commemorated by the churches and other cultural institutions in the years and decades following 1953. Before turning to the 1953 floods in more detail, I would like to reflect on the literature relating to cultural understandings of climate and weather, cultural memory, disaster studies and the commemoration of disaster in the United Kingdom.

Connecting conceptions of climate, cultural memory and disaster studies

Over the last decade or so, scholars across the humanities have begun to explore what climate and climatic change mean to individuals and specific communities (e.g. Raynor and Malone, 1998; Janković and Barboza, 2009; Geoghegan and Leyson, 2012). As this scholarship has developed, it has become apparent that a more detailed understanding of how individuals construct their conception of the climate, and how this is connected to their experiences of both everyday (de Vet, 2013) and exceptional weather (Hall and Endfield, 2016), is imperative to understanding how the public perceive climate change (Palutikof et al., 2004; Lorenzoni and Pidgeon, 2006). Further, as Hulme (2014) has demonstrated, the increasing trend to attribute individual extreme meteorological events to anthropogenic climate change raises many issues about the ambiguous meaning of causation. This challenges us to explore the relationship between scientific and technological understandings of global systems and everyday regional political and social realities (Hulme, 2014).

Historians have begun to show how cultural and social understandings of both past climates (Behringer, 2010) and historical meteorological extremes (Steinberg, 2006) can illuminate current debates and help to anchor contemporary literature on cultural dimensions of anthropogenic climate change within broader narratives of human-environment interaction. Studies have shown that the specific ways that everyday and extreme weather is remembered and recorded by a community influences both how future generations understand the risk posed to them by extremes and how individuals within the community perceive their climate and climatic change (Palutikof et al., 2004; Lorenzoni and Pidgeon, 2006; Hall and Endfield, 2016). Further, studies have begun to show how tangible commemorations, such as physical markers or newspaper features, may be bound up with less tangible and more emotive or nostalgic tendencies amongst individuals in a community (Gorman-Murray, 2010; Hall and Endfield, 2016: 14). Thus, for a more complete understanding of a community's cultural memory of a disaster, and how this may relate to present community resilience and social cohesion, one must attempt to explore how physical actions in the aftermath of an event interacted with individuals' wellbeing and understanding of events (e.g. Baxter, 2005).

Similar calls to embed meteorological extremes, and in turn catastrophes resulting from them, within broader studies of society and the everyday, have also emerged in the field of disaster studies (e.g. Blaikie et al., 2014: 330–346). In recent decades this multidisciplinary domain has grown substantially and now includes perspectives from economics (e.g. Merz et al., 2010), sociology (e.g. Rodriguez et al., 2009), anthropology (Hoffman and Oliver-Smith, 2002), history (Steinberg, 2006) and human geography (Blaikie et al., 2014).[3] These studies empirically investigate societal and community aspects of disasters, exploring communities' preparation, responses and resilience to floods, hurricanes and other extreme meteorological events. Such approaches have demonstrated that catastrophes are rarely caused by exceptional meteorological conditions alone.[4] The true risk posed by such a hazard is a combination of the probability of it occurring, the exposure of a community to the hazard and the vulnerability of the community to its impact (Diaz and Murnane, 2008: 12).[5]

As disasters appear to be influenced by a myriad of often-controllable factors, studies on the vulnerability of communities to specific hazards have resulted in the emergence of the applied systematic approach of disaster risk reduction (DRR). Included in the Millennium Development Goals set by the UN in 2000 (Blaikie et al., 2014: 325–327) and incorporated into their successor the Sustainable Development Goals of the 2030 Agenda for Sustainable Development (United Nations, General Assembly, 2015: 13–27), today DRR is used by many national governments, intra-governmental organisations including the United Nations Office for Disaster Risk Reduction[6] and non-governmental organisations, such as aid agencies. DRR aims to improve socioeconomic vulnerabilities to disaster, while reducing the risk posed by the environmental hazards that may trigger them. DRR aims to go beyond reactive emergency management, placing the improvement of a community's resilience to hazards at the heart of developmental, humanitarian and environmental programmes (UNISDR, 2016).

A key part of DRR is intra-agency planning, which integrates non-governmental organisations, local communities and other informal networks into disaster management strategies. As this approach has matured, over the last decade scholars have explored the specific role that non-governmental organisations have played in disaster. Empirical case studies have highlighted the role performed by such institutions and agencies in the immediate aftermath of disasters; when acting as part of an organised network (e.g. Moore et al., 2003) or when performing informally in response to a catastrophe (e.g. Airriess et al., 2008). These studies have explored the characteristics of such non-governmental organisations that allow them to play pivotal roles in disaster management: characteristics such as their level of integration within a community, the level of trust or authority they hold within an ethnic or socioeconomic group (Elliott and Pais, 2006), the speed with which they can respond, and their knowledge of vulnerable locations and individuals (Blaikie et al., 2014: 321–347). In short, these studies explore how community resilience in the aftermath of a disaster is connected to the community's preceding social cohesion and the strength of its social structures (Paton and Johnston, 2006).

Although religious groups and institutions often exhibit many of the above characteristics, deemed crucial for successful disaster management and post-disaster community resilience, they are almost completely absent from all major recent international treaties and scientific literature on the subject (Gaillard and Texier, 2010). As part of a wider collection attempting to address this gap in the literature, Wisner (2010) suggests that though little studied to date, faith groups do play an active role in the response, recovery, preparedness and prevention of disasters. Likewise, religious organisations are largely absent from historical narratives of disasters; where they do appear it is often in a moral or theological context (e.g. Steinberg, 2006: 12–19). An exception is Remes' *Disaster Citizenship* (2015), which for two technological disasters – the 1914 Salem fire and the 1917 Halifax explosion[7] – explores in detail the role local churches played in immediate disaster response and more long-term social support. For as Remes' states:

> People rarely record borrowing a cup of sugar from a neighbor...But when those same neighbors rescue each other after a disaster, people take notice and record the event. Disasters thus produce unusual records that document where people turned in times of trouble or crisis. Unions, churches, and mutual-aid societies...were not designed for disaster relief, but how they behaved in disasters shows us something about how they functioned.
>
> (Remes, 2015: 4)

Building on Remes' book, this chapter is an attempt to continue rectifying this gap in the literature, by exploring for one specific historical example how local churches provided vital social functions in the aftermath of a catastrophic flood.

Across communities up and down the length of the United Kingdom there are reminders of catastrophes and disasters past. From physical statues and markers at memorial sites (Williams, 2008: 8), through to folkloric traditions, such as songs, poems and stories (Stein and Preuss in Hartman, 2006; Griffiths et al., this volume) these reminders serve to commemorate the tragedy of lives lost, land and property damaged, community collapse and mistakes made. Whether physical or more ephemeral – referred to respectively as tangible and intangible cultural heritage (Boswell, 2008: 15) – such markers and reminders play an important role in intergenerational commemoration (Hall and Endfield, 2016) and collective memory (Barnier and Sutton, 2008).

The role that intangible cultural heritage plays in social and collective memory is not limited to the commemoration of disaster; many commemorations exist as part of wider folk narratives on a region, its communities or specific ethnic groups. Cultural commemoration of disaster is a central and ancient feature of many folkloric traditions, including myths – for example the Genesis flood narrative (*Good News Bible*, 1976: 9–13), literature – for example Daniel Defoe's *The Storm* (2005), poetry – for example the *Epic of Gilgamesh* (George, 2003) and folk songs – for examples see *Disaster Songs*, a catalogue of songs relating to disaster in Canada (Sparling, 2012). The North Sea Floods of 1953 have been encompassed into such traditions, with most recently the rock band *British Sea Power* releasing an ode to the floods' impact on the community of Canvey Island, Essex (Wilkinson, 2008). With the onset of modern printing, recording, digitising and media-sharing technologies, many of these previously oral traditions have now entered into more physical, but not necessarily more enduring, formats. Another important medium that often commemorates disasters are newspapers. Being geographically bound in coverage, often to a specific region or locale, and usually enduring for a substantial period of time, at least in generational terms, newspapers play an important role in commemorating events and generating regional narratives and histories (Zelizer and Tenenboim-Weinblatt, 2014).

More tangible cultural heritage in the form of statues, plaques and other markers commemorating memorial sites of naturally triggered disasters and their victims can be found all over the UK. Like the intangible formats discussed above, they perform an important role in intergenerational commemoration and collective memory, as well as a more direct function in land use and town planning. Being reminded of the height of previous flooding every time you encounter the marker or knowing that a specific plot of farmland has never been built on because it is susceptible to sink holes can save time, money and also lives (see for example Sargent, 1992: 14).

Often outlasting technocratic solutions, such as governmental records or flood maps,[8] simple flood markers placed at the high water mark of floods can be found hidden on village greens, community halls, churches and monuments all over the UK.[9] One specific flood marker, in the entrance hall of St Margaret's Church in King's Lynn, Norfolk, led me to begin thinking about the commemoration of disaster, alongside the empirical literature on community resilience and the role of local non-governmental organisations in disasters.

The marker records the high-water level for all flooding that has afflicted the church since the nineteenth century (Baxter, 2005: 1309), with the high water level from 1953 – the worst natural disaster in twentieth-century Britain (Hall, 2015: 2) – almost a foot below the mark of the less deadly and less well known 1978 floods (Figure 7.1). The flood markers at St Margaret's

Figure 7.1 Flood markers in the entrance to St Margaret's Church in King's Lynn,
     Norfolk England.
Photographs: author's own.

clearly highlight the importance of preserving specific local information,
reminding locals that the area surrounding this church has had water higher
than the national disaster of 1953, which claimed over 300 lives. Further, the
discrepancy between the 1978 marker and the lower 1953 one reminds us
that flood depth is only one of many factors that contribute to the severity of
a flood event, the damage it causes and in turn its commemoration.

Floodplain conditions surrounding the church have altered significantly
since it was first constructed in the early decades of the twelfth century
(Pevsner and Wilson, 2002: 460), with shops, homes and businesses surrounding the building having been built by almost every generation since. Upon visiting St Margaret's, I began to wonder how much the local community took
note of these flood markers. Had the church, itself at the centre of life in the
town both geographically and culturally, ever gone beyond the humble mark-
ing of flood levels to commemorate the lives lost in past floods and to help
embed these events within the cultural history and memory of the region? If
the church had played a role during past floods, such as those in 1953, how
had their role in the aftermath and years following the disaster interacted

with and affected the community's resilience to future floods? In essence, as John Urry asks, how does the act of commemorating or remembering effect the society within which it occurs and "are these processes of collective re-membering changing in the contemporary world…?" (Urry, 1995: 46)

Broadening out from St Margaret's church to include other local churches and religious organisations, I began by focusing my research on the role and function performed by these religious institutions during the North Sea floods of 1953.

The North Sea Floods of 1953 and their aftermath in King's Lynn

The North Sea Floods of 31 January and 1 February 1953 were caused by the combination of a large meteorological depression and record spring tides (Hall, 2011). The storm triggered a major storm surge, which flooded vast swathes of the east coast of Britain, the Netherlands and Belgium. In England alone it caused 1,200 breaches of sea defences resulting in over 160,000 acres of land being flooded, the evacuation of over 32,000 citizens, damage to 24,000 properties and 440 deaths (Steers, 1953; Baxter, 2005; Hall, 2015).[10] In Belgium it caused 22 deaths, and in the Netherlands, where the breaching of key dikes flooded vast areas of below sea-level polders, the storm surge resulted in 1,836 deaths (Gerritsen, 2005).

The progression of the storm surge southward along the east coast of England was relatively slow, making first landfall at Spurn Head, Yorkshire, at 16:00, yet not reaching communities further south, where most deaths oc-curred, until several hours later. For example, Canvey Island in Essex, where 58 deaths occurred, wasn't inundated until 01:10 on 1 February (Baxter, 2005: 1295). In spite of this significant lag time between the first landfall and the devastation caused further down the coast, no direct public warnings were issued, and each afflicted community had to deal independently with the deadly flood water (Hall, 2011: 389).

Much of the immediate rescue work was carried out by local authorities, communities, and military serviceman, both from the UK and US, based in the region (Baxter, 2005: 1300–1302). Although central government was initially slow to react, in the week that followed they declared the flooding a national disaster. Politicians and the national media were quick to invoke a wartime narrative of resilience and national solidarity (Furedi, 2007: 238; Hall, 2011: 390). Influenced by this, in the weeks that followed there was a national outpouring of support, with money, clothing and other sundries be-ing donated from all over the country. The invocation of wartime narratives was familiar to all British citizens in this post-war context of austerity, where rationing was still the norm and the infrastructure, including the very sea defences breached by the flood water, was still dilapidated (Hall, 2015: 3–4).

The storm surge reached its maximum at King's Lynn at 19:20 that evening, with its height at the town's harbour being calculated as somewhere between

2.5 (Steers, 1953: 285) and 2.7 metres above mean sea-level (Rossiter, 1954: 48). Approximately one-fifth of the town was inundated, around 1,800 homes were evacuated (Pollard, 1978: 44) and water as deep as 5.8 metres engulfed those properties closest to the main waterways (Map 7.2). Despite the widespread inundation, casualties were relatively low, with only 15 deaths being recorded in the town (Pollard, 1978: 44). Although the flooding occurred after dark, the fact that many people had not yet gone to bed prevented further lives from being lost, a pardon that was not afforded many communities further along the coast.

As the flood crippled communication lines along the east coast of Britain, and the national government's unpreparedness for such a catastrophe was exposed, each community had to rely on its own local networks. Local police, town councils and volunteers were central to immediate efforts to save lives, feed and shelter those who had spent the night exposed to the storm and ultimately to collect and deal with the dead (Baxter, 2005: 1300).[11] Despite the relatively low number of casualties in King's Lynn, the town is

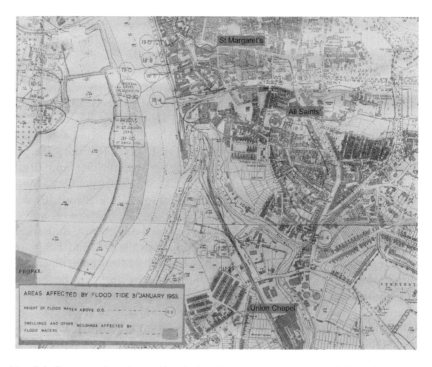

Map 7.2 County planning officer's hand-annotated map depicting the extent of the 1953 flooding in King's Lynn, with the churches – St Margaret's, Union Chapel, and All Saints – highlighted (depths shown are in feet). N.B. Stepney Church was located just to the north of St Margaret's, in an area not covered on this original 1953 map.

Source: NRO, C/P 3/2 (Used with kind permission).

the administrative centre for a large number of scattered rural communities in the region. Along the 15-mile coastal stretch from King's Lynn to Hunstanton alone there were a further 65 deaths (Baxter, 2005: 1298); the town became a hub for regional emergency response to the catastrophe.

Those churches in King's Lynn that hadn't been flooded quickly became de-facto emergency centres, offering shelter, warmth and food to those who had lost their homes. With St Margaret's in central King's Lynn still under nearly two feet of water,[12] the Union Chapel in South Lynn became "the principle haven of refuge"[13] (Map 7.2). Local churches were central to the collection and distribution of immediate aid and provisions, distributing food, clothes and other sundries across the afflicted communities.[14]

Exploring in detail the response of regional community organisations, such as the churches in King's Lynn, helps build a more detailed understanding of how the response to the floods was delivered and understood by those involved. Like Remes (2015), this approach allows us to explore the importance of an institution central to the community in 1953 and in following how the community commemorated the floods, to track whether this importance and centrality remained in the decades that followed.

Whilst national media and government in 1953 focused on the exceptional nature of the floods and the resilient wartime spirit with which communities responded (Furedi, 2007; Hall, 2011), the narrative that emerges at the community level in King's Lynn is one still centred on resilience, but also characterised by continuity and the normalisation of flooding in the region. Take, for example, the congregation of All Saints' church in King's Lynn (Map 7.2), 24 of whose members turned up for the service on the Sunday morning immediately after the overnight flooding. This is despite the church already being flooded to a depth of seven inches when the congregation had left Saturday Evensong. The Rector's home was also under three feet of water, but he celebrated Holy Eucharist at 11:00 and led the efforts of the congregation the following day to clean the church, so that it could be ready for the first funerals of victims on Wednesday 4 February 1953.[15] These events highlight the central role the church played in King's Lynn, both as part of the everyday routine and rhythm of the community and in times of disruption and dislocation. Through this one, simple anecdote we see the churches embedded and valued within the community, a characteristic heralded in disaster studies as strongly influencing community resilience and successful post-disaster recovery (Aldrich, 2012; Blaikie et al., 2014).

Alongside accounts of other non-governmental organisations such as the Red Cross,[16] the two local newspapers covered in detail the multiple roles local churches played in the immediate aftermath in King's Lynn. However, the church records themselves make very little reference to the floods. Where they do mention events of late January and early February 1953, the severity and exceptional human toll is noted, but across all of the meeting minutes and log books it is given at most a paragraph of discussion.[17] This is despite the fact that the committee for Stepney Baptist church, met only three days after the floods, and calmly noted in their meeting minutes that, "[t]he Minister and Church Secretary left after tea because of a flood warning for their area".[18] The two main Baptist churches, Union Chapel and Stepney, did produce a joint memorandum commemorating the floods, yet rather than circulating this amongst their congregations, they decided that the report should "be placed in the Minute Books of both Union and Stepney Churches to act as permanent memorial of the catastrophe".[19] These somewhat humble and understated gestures hint at the normalcy of flooding and the threat from the sea in the longer history of the town and these congregations. The sea and its dangers were never far away from life in the region, and as community figureheads local rectors were only too familiar with the risks posed to their coastal congregations, many of whom made their livelihoods from the sea. Only six years earlier, the Reverend Canon R. L. Whytehead, now of St Margaret's in King's Lynn, had led a service live on the BBC Home Service commemorating those from the region who had died at sea during the Second World War.[20]

In the weeks that followed the floods, the churches – along with other local volunteer non-governmental agencies such as the Scouts and Junior St John Ambulance Brigades (Pollard, 1978: 94) – continued to play a central role in relief efforts. All of the main congregations established their own disaster relief funds for those within their catchment and of their particular denomination. The church leadership promoted the plight of those in the region through national and international forums, including the Anglican national church assembly held in London[21] and BBC newsreel footage of the region, which featured coverage of a procession to a memorial service at All Saints' church.[22] The widespread coverage of the floods coupled with the large death toll meant that soon churches across the afflicted region were receiving donations from congregations as far afield as Durban in South Africa.[23]

In addition to their involvement with physical relief efforts, the churches played an important and cathartic social role in helping the communities to deal with the emotional and psychological trauma of the disaster. Recent sociological and psychological studies of disaster have shown that such support is integral to a community's social resilience and post-disaster recovery (Pitt, 2008: 357–366; Davis et al., 2015).

Although even as recently as the UK floods of 2007 there was criticism of a lack of psychological support provided for flood victims (Pitt, 2008: 357–366), services such as counselling and community support groups are now a formal part of disaster response and relief efforts in relation to flooding in the UK (Public Health England, 2014). Indeed, the foundation of much of the UK's disaster policy, planning and management today was first created in response to the 1953 floods (Waverley, 1954; Hall, 2011: 397–399). Yet in 1953 such co-ordinated and formalised provision of post flood support for trauma or bereavement was non-existent (Baxter, 2005: 1303–1305). As interviewees featured in a BBC documentary on the flooding reveal, the results of this internalised suffering can be profound, "typified by the two surviving Manser family members, who lost three siblings that fateful night, stating that they have lived their lives without closure because their parents never spoke of events, and they did not even know where their siblings were buried until fifty years after the flood" (BBC, 2002; Hall, 2011: 403).

Unfortunately, direct records of the informal emotional and psychological support local churches provided in lieu of more formal post-flood support are limited to the occasional quote from a memorial service or mention of support groups in local newspapers. In one such instance, the rector of All Saints' Church, the Reverend W.G. Bridge, reflected:

> Rescue work has been followed by efforts of rehabilitation, but through it all there has run a theme – the love of the community for all those who belong to it. That brotherly love has shown itself not least in our mourning for our dead.[24]

Through these limited glimpses in the archival record we can see that the social and community support role played by the churches in King's Lynn, although informal and ad hoc, was of great value to the community and ultimately seems to have increased social cohesion and resilience in the aftermath of the tragedy. For community members, the church was a space that combined a sense of belonging tied to a specific place, with personal faith, which in the aftermath of the floods proved important for both physical and emotional resilience.

Commemorating the 1953 floods in King's Lynn

All of the churches in King's Lynn continued to play a central role in funerals, memorial services and the distribution of relief in the months that followed the flooding. Yet beyond the individual families who lost loved ones, for the wider community the floods quickly became another flood from the past, an accepted event in the region's history. There is no evidence in either the local press or church log books that the floods were commemorated on their one, two or five year anniversaries. In fact, the three churches log books I have been able to locate contain no mention of any commemoration ceremonies or services after 1953. However, it is worth noting that these log books do not document the content of every individual sermon, and so we can perhaps speculate that some of the reverends in King's Lynn may have dedicated a prayer to victims on these initial anniversaries.

From the archival record, however, it is clear that in the initial months and years that followed the floods, while communities were vociferous about the failings of infrastructural and government response (Pollard, 1978: 93–106; Hall, 2012: 124–166), there was also a desire and urgency to return to normal as quickly as possible. Throughout 1953 and into 1954, many victims faced problems with insurers paying out for damages, slow progress on repairing homes due to post-war shortages and living in temporary accommodation for more than six months after the floods (Pollard, 1978: 96–100).

Contemporary studies have shown how social capital – "the networks and resources available to people through their connections to others" – plays a vital function in post-disaster recovery (Aldrich, 2012: 2). Although I do not wish to directly and anachronistically apply this modern concept to events of 1953, the disparities in post-disaster recovery and their relation to individuals' access to social resources, social networks and social capital evident in modern case studies (e.g. Aldrich, 2012) echoes disparities evident in 1953. Disparities between recovery in towns such as King's Lynn and smaller rural communities, between middle-class and working class victims, and between those integrated into community networks, such as church congregations, and those living on the margins are all evident and remarked upon in accounts of survivors (Pollard, 1978; BBC Radio Norfolk, 1993; BBC, 2002).

Although victims were commemorated with physical markers, statues and plaques in nearly every afflicted town or village along the north Norfolk coast, beyond 1953 other forms of commemoration are not evident from the region's archival records. The earliest evidence of widespread commemoration appears in 1973 on the 20th anniversary of the floods and is led largely by the local and regional press. Most notably, the *Eastern Daily Press*, a title covering the whole county of Norfolk, produced a four-page commemoration supplement on the 1973 anniversary, which was much more critical of events in 1953 than the contemporary media coverage had been.

From our vantage point in the present, the gap between the floods and the first press commemoration may seem like a relatively long time, but we must remember that by the 1970s many aspects of the floods were still fresh and emotionally raw for those still living and working in the region. As Pollard reflected in 1977:

> Many East Anglian families, and not only those who lost loved ones, are still psychologically marred by the disaster. For some there are occasional physical reminders, as on one Norfolk farm where the plough still, nearly twenty-five years later, occasionally turns up pots of ink, bottles, combs and other items from…the sea…A continuing souvenir for thousands of home-owners in the coastal towns and villages is the difficulty of decorating walls which were saturated with sea-water and will, according to expert advice, never again take and hold paint or paper satisfactorily.
>
> (Pollard, 1978: 8)

The 20-year gap between the floods and their commemoration in the regional press may reflect the importance of oral traditions and intergenerational narratives in the commemoration of extreme weather events (Hall and Endfield, 2016). The interactions and dynamics of individual memories, collective memories and longer more formalised histories of a region, which may result in an event's being commemorated or forgotten, are indeed multiple, bespoke and complicated.[25]

Five years later, the 25th anniversary received significant press coverage as it occurred just weeks after the 1978 floods, which again subjected the region to extensive coastal flooding. As highlighted by the flood markers at St Margaret's church (Figure 7.1), although less widespread in their devastation, the 1978 floods did cause significant damage in King's Lynn and several other locations along the north Norfolk coast.[26] Nationally, the 1978 floods were hailed as a success for the changes made post-1953 to warning systems, sea defences and disaster planning, but regionally, there was still plenty of criticism of the government's response (see Hall, 2011: 394–404). The coverage of the 1978 flooding and the 25th anniversary of 1953 highlight how socioeconomic and political circumstances had changed in King's Lynn during the intervening years. Whilst the damage to the physical building of St Margaret's Church and its clean-up were covered in the regional press,[27] there was no mention of commemorative services, and reflecting wider trends of secularisation, the church hierarchy no longer had a presence in print, as the "Church News and Views section" in the *Lynn News & Advertiser* had been discontinued.

Further, the 25th anniversary was used by the newspapers as an opportunity to criticise both the local and national government's preparedness and response to the 1978 floods.[28] The most telling example that the 25th anniversary occurred in a more secular, individualistic and relatively prosperous period is the response to the local Salvation Army's relief efforts; just as in 1953 the Salvation Army in King's Lynn collected furniture and clothing for those in need. However, unlike in 1953, when people queued to receive the donations, in 1978 no one came to collect any of the donated items.[29] With the organisation left with a hall full of donated shoes, clothes and furniture, the organiser felt aggrieved when local communities complained that no one was helping them. He lamented: "People seem to want a cash settlement. But we have no money to hand out. We have just done what we can to help the problem as we understood it".[30]

The centrality and importance of the role played by the churches in King's Lynn in the recovery from the 1953 floods becomes much more apparent when it is considered alongside 1978. The different and often more marginal role the church played in the aftermath of the 1978 floods reflects studies on secularisation in the UK, which identify the 1960s as a key decade for the receding importance of the church across many areas of British life (Brown, 2013: 170–192). Rather than simply being related to a purported general secularisation of British society in the period, most often measured in declining church attendance and religious adherence figures (e.g. Voas and Bruce, 2004; Voas and Crockett, 2005; Voas, 2009), the different role the church played in the aftermath of the 1978 floods in King's Lynn perhaps reflects more nuanced changes in the town's social structures. Green (2011) argues for a more socially informed history of religion, demonstrating the importance of integrating church trends within their myriad social influences, which most notably in the post-war decades includes the expansion of the welfare state. In the context of the 1953 and 1978 floods in King's Lynn, the expansion of organised state services is observed through the creation of both national and regional weather warning systems and governmental disaster plans in the aftermath of the 1953 catastrophe (Johnson et al., 2005; Hall, 2015). In providing this disaster-specific support, along with more general increases in social welfare, the state took up many of the responsibilities that in the immediate aftermath in 1953 had been picked up by the local churches. The changing role of the churches in King's Lynn in this period highlights not only a local shift in community identity and its consequences for social cohesion and resilience, but also reflects larger changes across Britain relating to secularisation, tradition and the stability of regional identity.

Whilst the churches were largely absent from the documented longer-term commemoration of floods, their role in the immediate aftermath must be considered as part of a process that embedded the floods into the region's collective memory and regional history. Beyond the tangible heritage created by the flood markers, we have seen that the churches played a central role in the intangible cultural heritage of the floods, contributing to their position within a broader collective memory and history of the region.

Conclusion

Whilst perhaps frustrating for academic enquiry, the limited and sporadic archival evidence of the churches' role in the aftermath and commemoration of the 1953 floods in King's Lynn is in itself important. The lack of records shows that what the church was doing, even during this time of disruption and catastrophe, was a normal role for it to undertake within the community. In a town and wider region whose history has been defined by water and the sea, including repeated flooding, the role of the churches in 1953 was essential and yet literally unremarkable.

We have seen that by being supportive and resolute and avoiding the superlatives of the media coverage, whilst also providing a forum for the expression of emotions, the churches and their associated community groups helped to normalise the flood and situate it within a longer history of repeated flooding in this low-lying region. By performing this function during the immediate aftermath, the church helped to situate the 1953 floods within the region's historical narrative, helping to ensure future commemoration would occur and that this commemoration would not just focus on the tragedy of the flood, but rather present a narrative that centred on the overcoming of adversity, social resilience and community social values. Given the ongoing success of the town of King's Lynn and the limited damage from flooding in the post-1953 era, this case study supports the idea that the incorporation of extreme weather events into a community's long-term social memory is important for the community's perception of future risks, adaptive development of infrastructure and ongoing social resilience.

Whilst there is evidence of how the church affected the intergenerational memory and commemoration of events, the difference in their role between the 1953 and 1978 floods has highlighted the challenges of exploring intergenerational and collective memory. The shift of the churches from institutions at the centre of support networks in 1953, to more peripheral actors in 1978, reminds us that history is not a static scene from which we can cherry pick case studies. Any serious study attempting to understand how past extremes of weather have been survived, understood and commemorated by communities must place events and actors within broader cultural and social histories (Remes, 2015: 4). Given the changing role of the church over much of the UK during the twentieth century, it is further suggested that future historical studies of disaster of this period should continue to explore the churches' role. In this instance, the church has provided a means of capturing the flux of social "networks of solidarity and obligation" (Remes, 2015: 4) so central to contemporary studies on social resilience, yet so often absent from the historical record.

Events in King's Lynn in the aftermath of the 1953 floods have highlighted that a cultural memory of extreme weather and disaster events may only endure through the complex combination of various mechanisms across a variety of mediums. We have seen that for this case study an intergenerational narrative, specific to the town and region, was only formed through a mixture of oral narratives, regional newspapers, physical markers and importantly the actions of community networks led by non-governmental organisations, in this instance the church.

Although occurring in a time before widespread acceptance of the threats posed by anthropogenic climate change, the narrative presented still informs us about how we should approach meteorological extremes and their commemoration within a community. In line with other recent literature (Hulme, 2009; Hall and Endfield, 2016), this chapter reminds us that when interacting with communities in relation to climate change, we should avoid the abstract and not under-value the importance of local understanding, social memory and experience of previous meteorological extremes.

Notes

1 Notable exceptions to this approach are Steinberg's (2006) *Acts of God: The Unnatural History of Natural Disaster in America*, Remes' (2015) *Disaster Citizenship: Survivors, Solidarity and Power in the Progressive Era* and *Mapping Vulnerability: Disasters, Development and People* by Bankoff et al. (2013).
2 Recent studies that highlight the importance of considering the broader societal contexts when presenting case studies on historical disasters include Morgan (in this volume) and Veale and Endfield (2016).

3 For an overview of disaster research across a range of disciplines see Quarantelli (1998).
4 For more see chapters by Morgan and Waites in this volume.
5 For more on this see Bankoff et al. (2013).
6 The United Nations Office for Disaster Risk Reduction was founded in 1999, for more information see www.unisdr.org/who-we-are/mandate (accessed 3 July 2016).
7 The Great Salem Fire of 25 June 1914 was caused by an explosion at a leather company and went on to destroy 1,376 buildings, making some 18,380 residents of Salem, Massachusetts, homeless, jobless or both (Remes, 2015: 54). The Halifax Explosion occurred on 6 December 1917, when a French cargo ship carrying explosives collided with a Norwegian steamer causing a fire and then an explosion, thought to be the largest man-made explosion before the atomic bomb, killing 2,000 people (Remes, 2015: 21–22).
8 For example, the oldest flood marker in Rome dates back to 1277 (Aldrete, 2007, Appendix I).
9 For more on coastal flood markers please see the UK Coastal Floodstone Project at http://floodstones.co.uk/. The project is developing a database of UK flood markers to which readers can contribute (accessed 15 July 2016).
10 Most scholarly accounts of the flood report total deaths in the UK of 307. However this total doesn't include the 133 deaths caused by the sinking of the MV *Princess Victoria*, which was caused by the same weather system. The true total death toll for the UK may be larger still, as it is not clear whether official statistics include several fishing vessels and their crews lost in the storm (Hall, 2015: 19).
11 For more on the development of UK disaster communications in response to the 1953 floods, specifically the development of a national coastal flood warning system see Hall (2012, 2015).
12 "The vicar looks back." *Lynn News & Advertiser*, 6 March 1953: 7 [Print].
13 Butler, F.W.J. "Church news and views." *Lynn News & Advertiser*, 24 February 1953: 7 [Print].
14 "Queues in a chapel for food and clothes." *Lynn News & Advertiser*, 6 February 1953: 12 [Print].
15 "Service held at All Saints' next morning." *Lynn News & Advertiser*, 6 February 1953: 8 [Print].
16 "Red cross to the rescue." *Lynn News & Advertiser*, 6 February 1953: 10 [Print].
17 See NRO FC 26/2, FC 69/24 and FC 65/9.
18 Minutes of meeting held 3rd February 1953. *Union Baptist Church Minute Book, 1909–1967.* [minutes] Records of King's Lynn Stepney Baptist, NRO FC 65/9.
19 Butler, F.W.J. (1953) *The Flood of January 31st, 1953.* [memorandum] Records of King's Lynn Union Baptist, NRO FC 26/2.
20 Whytehead, R.L. (1947) "Harvest of the sea." 19 January 1947. *Radio Times* [online] BBC Genome. http://genome.ch.bbc.co.uk/809ab8cf05c84ba1af9070760341fabc (accessed 16 July 2016).
21 "Lynn Vicar's Church assembly plea for flood sufferers." *Lynn News & Advertiser*, 17 February 1953: 7 [Print].
22 "Magdalen Lynn on television." *Lynn News & Advertiser*, 17 February 1953: 5 [Print].
23 Wellington, W.L. (1953) Letter to the Rector St John's Church. [letter] File of correspondence relating to flood damage, mainly to the church, NRO PD 171/82.
24 Bridge, W.G. (1953) Quoted in: "Memorial service at South Lynn church." *Lynn News & Advertiser*, 10 February 1953: 1, 5 [Print].
25 For more on the complexity of this relationship, see the 2008 special edition of the journal *Memory*, "From individual to collective memory: theoretical and

empirical perspectives," edited by Barnier and Sutton. www.tandfonline.com/toc/pmem20/16/3 (accessed 12 July 2016).
26 For largely comparable accounts of the meteorological and hydrographical conditions in both 1953 and 1978 see Steers (1953) and Steers et al. (1979).
27 E.g. "Church clean-up." *Lynn News & Advertiser*, 27 January 1978: 2 [Print].
28 "25th anniversary of flooding threatened nightmare repeat." *Eastern Daily Press*, 30 January 1978: 7 [Print].
29 "Salvation Army wants to help." *Lynn News & Advertiser*, 3 February 1978: 2 [Print].
30 "Lynn victims' second chance for SA help." *Eastern Daily Press*, 2 February 1978: 18 [Print].

References

Airriess CA, Li W, Leong KJ, Chen ACC and Keith VM (2008) Church-based social capital, networks and geographical scale: Katrina evacuation, relocation, and recovery in a New Orleans Vietnamese American Community. *Geoforum* 39 (3): 1333–1346. doi:10.1016/j.geoforum.2007.11.003.

Aldrete GS (2007) *Floods of the Tiber in Ancient Rome*. Baltimore, MD: JHU Press.

Aldrich DP (2012) *Building Resilience: Social Capital in Post-Disaster Recovery*. Chicago, IL: University of Chicago Press.

Bankoff G, Frerks G and Hilhorst D (2013) *Mapping Vulnerability: Disasters, Development and People*. Oxford: Earthscan.

Barnier AJ and Sutton J (2008) From individual to collective memory: theoretical and empirical perspectives. *Memory* 16 (3): 177–182. doi:10.1080/09541440701828274.

Baxter PJ (2005) The east coast Big Flood, 31 January–1 February 1953: a summary of the human disaster. *Philosophical Transactions. Series A, Mathematical, Physical, and Engineering Sciences* 363 (1831): 1293–1312. doi:10.1098/rsta.2005.1569.

BBC (2002) *Timewatch: The Greatest Storm*. [TV Programme] www.youtube.com/watch?v=vARjm3yHKzY (accessed 12 July 2016).

BBC Radio Norfolk (1993) *A Flood of Memories*. [Radio Programme] Norfolk Sound Archive, Norfolk Record Office, SAC 2003/3/425, Norwich.

Behringer W (2010) *A Cultural History of Climate*. Cambridge: Polity.

Blaikie P, Cannon T, Davis I and Wisner B (2014) *At Risk: Natural Hazards, People's Vulnerability and Disasters*. Oxford: Routledge.

Boswell R (2008) *Challenges to Identifying and Managing Intangible Cultural Heritage in Mauritius, Zanzibar and Seychelles*. Oxford: African Books Collective.

Brown CG (2013) *The Death of Christian Britain*. Oxford: Routledge.

Davis A, Davis C and Gibson J (2015) *Psychological Support in Disasters: Facilitating Psychological Support for Catastrophic Events*. San Clemente, CA: LawTech Publishing Group.

de Vet E (2013) Exploring weather-related experiences and practices: examining methodological approaches. *Area* 45 (2): 198–206. doi:10.1111/area.12019.

Defoe D (2005) *The Storm*. London: Penguin (Reprint of 1703 original).

Diaz HF and Murnane RJ (2008) *Climate Extremes and Society*. Cambridge: Cambridge University Press.

Elliott JR and Pais J (2006) Race, class, and hurricane Katrina: social differences in human responses to disaster. *Social Science Research*, Katrina in New Orleans/

Special Issue on Contemporary Research on the Family 35 (2): 295–321. doi:10.1016/j.ssresearch.2006.02.003.

Furedi F (2007) From the narrative of the blitz to the rhetoric of vulnerability. *Cultural Sociology* 1 (2): 235–254. doi:10.1177/1749975507078189.

Gaillard JC and Texier P (2010) Religions, natural hazards, and disasters: an introduction. *Religion* 40 (2): 81–84. doi:10.1016/j.religion.2009.12.001.

Geoghegan H and Leyson C (2012) On climate change and cultural geography: farming on the Lizard Peninsula, Cornwall, UK. *Climatic Change* 113 (1): 55–66. doi:10.1007/s10584-012-0417-5.

George A (2003) *Epic of Gilgamesh*. London: Penguin.

Gerritsen H (2005) What happened in 1953? The big flood in the Netherlands in retrospect. *Philosophical Transactions: Mathematical, Physical and Engineering Sciences* 363 (1831): 1271–1291.

Good News Bible (1976) Swindon: The Bible Societies/Collins Bible.

Gorman-Murray A (2010) An Australian feeling for snow: towards understanding cultural and emotional dimensions of climate change. *Cultural Studies Review* 16 (1): 60. doi:10.5130/csr.v16i1.1449.

Green SJD (2011) *The Passing of Protestant England: Secularisation and Social Change, C. 1920–1960*. Cambridge: Cambridge University Press.

Hall A (2011) The rise of blame and recreancy in the United Kingdom: a cultural, political and scientific autopsy of the North Sea Flood of 1953. *Environment and History* 17 (3): 379–408. doi:10.3197/096734011X13077054787145.

Hall A (2012) Risk, blame, and expertise: the meteorological office and extreme weather in Post-War Britain. Manchester eScholar Services, University of Manchester, Manchester. www.escholar.manchester.ac.uk/item/?pid=uk-ac-man-scw:181772.

Hall A (2015) Plugging the gaps: the North Sea Flood of 1953 and the creation of a National Coastal Warning System. *Journal of Public Management and Social Policy* 22 (2). http://digitalscholarship.tsu.edu/jpmsp/vol22/iss2/8.

Hall A and Endfield GH (2016) "Snow scenes": exploring the role of memory and place in commemorating extreme winters. *Weather, Climate, and Society* 8 (1): 5–19. doi:10.1175/WCAS-D-15-0028.1.

Hartman CW (2006) *There Is No Such Thing as a Natural Disaster: Race, Class, and Hurricane Katrina*. London: Routledge.

Hoffman S and Oliver-Smith A (2002) *Catastrophe & Culture: The Anthropology of Disaster*. Santa Fe, NM: School of American Research Press.

Hulme M (2009) *Why We Disagree about Climate Change: Understanding Controversy, Inaction and Opportunity*. Cambridge: Cambridge University Press.

Hulme M (2014) Attributing weather extremes to "climate change": a review. *Progress in Physical Geography*, 38 (4): 499–511. doi:10.1177/0309133314538644.

Hunter-Blair A (2001) *River Great Ouse and It's Tributaries*. Cambridgeshire: Imray, Laurie, Norie & Wilson, Limited.

Janković V and Barboza CH (2009) *Weather, Local Knowledge and Everyday Life: Issues in Integrated Climate Studies*. Rio de Janeiro: MAST.

Johnson CL, Tunstall SM and Penning-Rowsell EC (2005) Floods as catalysts for policy change: historical lessons from England and Wales. *International Journal of Water Resources Development* 21 (4): 561–575. doi:10.1080/07900620500258133.

Knittl MA (2007) The design for the initial drainage of the Great Level of the Fens: an historical whodunit in three parts. *The Agricultural History Review* 55 (1): 23–50.

Lamb H and Frydendahl K (1991) *Historic Storms of the North Sea, British Isles and Northwest Europe*. Cambridge: Cambridge University Press.

Lorenzoni I and Pidgeon NF (2006) Public views on climate change: European and USA perspectives. *Climatic Change* 77 (1–2): 73–95. doi:10.1007/s10584-006-9072-z.

Merz B, Kreibich H, Schwarze R and Thieken A (2010) Assessment of economic flood damage. *Natural Hazards and Earth System Sciences* 10 (8): 1697–1724. doi:10.5194/nhess-10-1697-2010.

Moore S, Eng E and Daniel M (2003) International NGOs and the role of network centrality in humanitarian aid operations: a case study of coordination during the 2000 Mozambique floods. *Disasters* 27 (4): 305–318. doi:10.1111/j.0361-3666.2003.00305.x.

Palutikof JP, Agnew MD and Hoar MR (2004) Public perceptions of unusually warm weather in the UK: impacts, responses and adaptations. *Climate Research* 26 (1): 43–59. doi:10.3354/cr026043.

Paton D and Johnston DM (2006) *Disaster Resilience: An Integrated Approach*. Springfield, IL: Charles C Thomas Publisher.

Pevsner N and Wilson B (2002) *Norfolk 2: North-West and South*. New Haven, CT: Yale University Press.

Pitt M (2008) *Learning Lessons from the 2007 Floods*. London: Cabinet Office.

Pollard M (1978) *North Sea Surge: The Story of the East Coast Floods of 1953*. Sudbury: Terence Dalton Limited.

Public Health England (2014) *Health Advice: General Information about Mental Health following Floods*. London: Public Health England. www.gov.uk/government/uploads/system/uploads/attachment_data/file/483387/Health_advice_about_mental_health_following_floods_2015.pdf (accessed 12 July 2016).

Quarantelli EL (1998) *What Is a Disaster?: Perspectives on the Question*. Oxford: Psychology Press.

Raynor S and Malone E (1998) Human choice and climate change. Volume 1: The societal framework. http://inis.iaea.org/Search/search.aspx?orig_q=RN:30038462.

Reilly B (2009) *Disaster and Human History: Case Studies in Nature, Society and Catastrophe*. Jefferson, NC: McFarland.

Remes JA (2015) *Disaster Citizenship: Survivors, Solidarity, and Power in the Progressive Era*. Champaign: University of Illinois Press.

Richards W (1812) *The History of Lynn: Civil, Ecclesiastical, Political, Commercial, Biographical, Municipal, and Military, from the Earliest Accounts to the Present Time*. London: Printed by W. G. Whittingham.

Rodriguez HE, Quarantelli EL and Dynes R (2009) *Handbook of Disaster Research*. New York: Springer Science & Business Media.

Rossiter JR (1954) The North Sea storm surge of 31 January and 1 February 1953. *Philosophical Transactions of the Royal Society of London A: Mathematical, Physical and Engineering Sciences* 246 (915): 371–400. doi:10.1098/rsta.1954.0002.

Sargent DM (1992) Flood management in Rockhampton, Australia. In Saul AJ (ed.) *Floods and Flood Management*, Springer: Netherlands: 3–15.

Sparling H (2012) Disaster songs. http://disastersongs.ca/ (accessed 12 July 2016).

Steers JA (1953) The East coast floods. *The Geographical Journal* 119 (3): 280–295. doi:10.2307/1790640.

Steers JA, Stoddart DR, Bayliss-Smith TP, Spencer T and Durbidge PM (1979) The storm surge of 11 January 1978 on the East coast of England. *The Geographical Journal* 145 (2): 192–205. doi:10.2307/634386.

Steinberg T (2006) *Acts of God: The Unnatural History of Natural Disaster in America*. Oxford: Oxford University Press.

UNISDR (2016) What is disaster risk reduction? www.unisdr.org/who-we-are/what-is-drr (accessed 3 July 2016).

United Nations, General Assembly (2015) *Transforming our world: the 2030 Agenda for Sustainable Development*, A/RES/70/1 (21 October 2015). http://undocs.org/A/RES/70/1 (accessed 10 August 2016).

Urry J (1995) How societies remember the past. *The Sociological Review* 43 (S1): 45–65. doi:10.1111/j.1467-954X.1995.tb03424.x.

Veale L and Endfield G (2016) Situating 1816, the 'year without summer', in the UK. *The Geographical Journal*. doi:10.1111/geoj.12191.

Voas D (2009) The rise and fall of fuzzy fidelity in Europe. *European Sociological Review* 25 (2): 155–168. doi:10.1093/esr/jcn044.

Voas D and Bruce S (2004) Research note: the 2001 census and Christian identification in Britain. *Journal of Contemporary Religion* 19 (1): 23–28. doi:10.1080/1353790032000165087.

Voas D and Crockett A (2005) Religion in Britain: neither believing nor belonging. *Sociology* 39 (1): 11–28. doi:10.1177/0038038505048998.

Waverley JA (1954) *Report of the Departmental Committee on Coastal Flooding*. London: HMSO.

Wilkinson JS (2008) *Canvey Island*. YouTube. www.youtube.com/watch?v=lAPipGclVk4 (accessed 12 July 2016).

Williams P (2008) *Memorial Museums: The Global Rush to Commemorate Atrocities*. London: Bloomsbury Academic.

Wisner B (2010) Untapped potential of the world's religious communities for disaster reduction in an age of accelerated climate change: an epilogue & prologue. *Religion* 40 (2): 128–131. doi:10.1016/j.religion.2009.12.006.

Zelizer B and Tenenboim-Weinblatt K (2014) Journalism's memory work. In Zelizer B and Tenenboim-Weinblatt K (eds.) *Journalism and Memory*. London: Palgrave Macmillan UK: 1–14.

8 "The ice shards are gone"

Traditional ecological knowledge
of climate and culture among the Cree
of the Eastern James Bay, Canada

Marie-Jeanne S. Royer

Introduction

> The people [...] they need real food. I remember some people, they re-
> ceive a few sickness and they really want to eat the real food. And when
> they eat the real food, in two days, the pain is gone. (Sic)
>
> (Anonymous, 2010)

Traditional Ecological Knowledge (TEK) is the knowledge that comes
about through continuous long- term contact with the environment that
surrounds us. Through this contact, humans learn about their environment,
its particularities and its cycles. In trying to classify what TEK is, Usher
(2000) separates it into four parts. The first revolves around factual and ra-
tional knowledge about the environment typically based on current or past
observations. The second centres on factual knowledge about past and current use of the environment. Here, cultural and personal memory play key
roles as the factual knowledge can be based on a variety of sources including
personal experience, past observations and oral history. The third part is
composed of culturally based value statements. Again, these can be influenced by a society's past interactions, observations and memories linked
to the environment. The fourth and final part underlies the first three and
revolves around culturally based cosmology. In modern society, it is easy
to see human influence on the environment, be it through species selection
and breeding or infrastructure and management systems. However, climate
and environmental conditions also influence culture and society. They have
been shown to affect language, beliefs, traditions and cultural/social institutions. These influences are sometimes easier to observe in communities
that continue to have direct interactions with the environment through the
practice of subsistence activities, such as many of Canada's northern Indigenous communities. For these communities, the continued practice of
traditional subsistence activities is intrinsically linked to cultural identity
and has extensive repercussions on health and well-being. Although Cree
lifestyle in the Eastern James Bay (see Map 8.1) has changed dramatically
since the middle of the twentieth century, the Cree still rely heavily on the

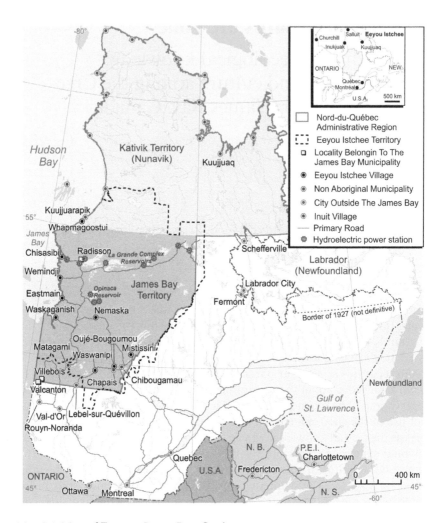

Map 8.1 Map of Eastern James Bay, Quebec.

local environment for many subsistence activities. As such, traditional ecological knowledge and the subsistence activities they relate to extends to all
aspects of Cree cultural and social life.

The success of subsistence activities relies not only on the practitioner's
hunting, fishing or trapping abilities but also on his understanding and knowledge of the environment and particularly weather conditions. Although TEK
is based on a long history of use of the land and natural resources from multiple generations, Cree, like many other aboriginal groups, are often wary of
generalising events beyond the scope of their own personal experience. They
will often distinguish personal observations from events they have heard

about but have not personally observed. This particularity is best illustrated through the quote from one of the Cree witnesses during the court proceedings over the development of the James Bay area. When asked to swear to tell the truth, he answered through his translator that "He does not know whether he can tell the truth. He can tell only what he knows" (Richardson, 1976). It is partially for this reason that intergenerational transmission of TEK and with it the transmission of cultural memory including cultural memory of weather, is often achieved through the practice of activities. It is passed on through actions more than words.

This knowledge of the weather and the environment can influence the success of a hunt but also a hunter's safety. Changes in the climate and local weather conditions can thus have far-reaching implications. The impacts of changes in local weather and environmental conditions can be direct, such as storms, or indirect, for example, by causing changes in the availability and location of resources or presence of parasites. The Arctic and sub-Arctic regions are believed to be among the first areas to feel the major impacts of climate change (Symon et al., 2005; Trenberth et al., 2007). Already, Cree populations have observed changes in weather patterns and environmental conditions. This chapter draws on TEK and scientific data to obtain a holistic vision of the changes occurring on the territory. It combines aspects from the sciences and the humanities in the hopes of reaching a clearer understanding of the potential impacts of these changes on the local Cree populations.

The Cree and social change

With an area of 860,553 km^2, the Nord-du-Québec Administrative Region (Québec, Canada) is the largest administrative region of Quebec representing 55% of the territory (MDDELCC, 2016). It is also its least populated with 0.5% (41,503 inhabitants) of the province's population living there in 2005. It is divided into the regional administration territories of the James Bay and Kativik. The James Bay Territory, or Eeyou Istchee in Cree, is the southern part of the region and covers 297,330 km^2, extending from 49°N to 55°N and from the James Bay at its western-most border to 68°W at its eastern-most border (Ducruc et al., 1976; MBJ, 2010; CRBJ, 2011). The James Bay Territory has five Quebec municipalities with 15,351 inhabitants (MAMROT, 2011). These communities were originally founded between 1935 and 1974 and were for the most part linked to the mining and hydroelectricity trades in the area. The Cree have nine villages on the territory with a total population of 16,350 inhabitants: Chisasibi, Eastmain, Mistissini, Nemaska, Oujé-Bougoumou, Waskaganish, Waswanipi, Wemindji and Whapmagoostui (Statistics Canada, 2012a–i). Each of these communities is a Cree Nation and has its own governments, called the Local Government of the First Nation, which manages the village and lands attached to it.

The Grand Council of the Cree (GCC) is the regional Cree government. This was created in 1974 during the negotiations for the James Bay hydro-electric project.

The history of the Eastern James Bay region is relatively 'quiet'. The first traces of occupation date back some 3,500 years (Feit, 1995; Morantz, 2002; Desbiens, 2004). At the arrival of Europeans in North America in the seventeenth century, the Cree First Nation of the Eastern James Bay were documented as living in family units of hunter-gatherers and moving along the vast territory divided into family hunting territories (Morantz, 1978; Tanner, 1983; Peloquin and Berkes, 2009). It is this long occupation of the Eastern James Bay and the practice of traditional subsistence activities that have allowed the Cree to develop extensive TEK of the area. Although the fur trade between the Cree and the Europeans, more particularly through dealings with the Hudson's Bay Company (HBC), became an important economic activity from 1670 onwards, traditional subsistence activities remained the basis of the Cree economy and its land management systems (Chaplier, 2006; SAA, 2009). Nevertheless, the network of trading posts the company established did have an impact on Cree social structure, as over time some members of the community started settling down near these posts (Duhaime, 2001). For scholars of the region, the Hudson Bay Company's ship and trading post log books remain the only written sources about people, weather, and environmental conditions in the area from that period (Catchpole, 1992). In 1870, the Eastern James Bay was transferred to the Dominion of Canada after the Hudson's Bay Company surrendered its charter to the British Crown. Not until 1987 did the Administrative Region take on its current shape and size. History doesn't seem to show that over this time either the Hudson's Bay Company or the governments of Quebec and Canada after confederation exerted any administrative or management pressure on the Cree territories of the James Bay Territory (Scott, 1988, 1989; Canobbio, 2009). This is best reflected by the statement of M. Ciacca in his foreword of the James Bay and Northern Québec agreement in 1975: "Until now, the native peoples have lived legally speaking, in a kind of limbo. The limits of federal responsibility were never quite clear, nor was it quite clear that Québec had any effective jurisdiction" (SAA, 1998: XV). Between 1867 and 1960, the area was considered under the legal and political management of the federal government and during this time administration was limited to a small number of nursing stations and schools, along with, in the twentieth century, annual visits (lasting a few days) by doctors and Royal Canadian Mounted Police (RCMP) officers (Niezen, 1993; Goudreau, 2003).

This attitude changes in the 1970s when the provincial government set its sights on developing a large-scale hydroelectric complex in the northern part of its territory. Following the Oil Crisis of the 1970s and with the United States' high demands in electric energy, many Canadian provinces believed that building mega-dams was the energy strategy of the future (White, 1995; Froschauer, 1999; Swyngedouw, 2007). This northern region was selected for the project as it held a special appeal as both a symbolic and real representation of the nature and resources of the province that fitted into the population's collective narrative (Hogue et al., 1979; Desbiens, 2004). The announcement of the project in 1971 led to the mobilisation of the Cree First Nation of the Eeyou Istchee who hadn't been consulted and to their appealing this project in front of courts of law (Chaplier, 2006; Savard, 2009). These actions culminated in 1975 with the signing of the James Bay and Northern Quebec Agreement (JBNQA), the first agreement between the governments of Canada and Quebec and members of Cree and Inuit representatives of the Nord-du-Québec Administrative Region. It represents the first modern treaty on global land claims in Quebec (CCRCCN, 1991). Because the JBNQA refers to land claims as defined by the constitution law of 1982, all institutions put in place through clauses from the JBNQA are constitutionally protected. The JBNQA was all encompassing including clauses related to environmental protection, justice administration, socioeconomic development, daily life and local and regional governments (CCRCCN, 1991). Due to the complex nature of the agreement, this was not the last word on the subject, and further agreements and amendments followed looking to solve many of the JBNQA's conflictual elements, the last one signed in 2013.

Until the JBNQA, the Cree lived in small, isolated communities that rarely had any contact with one another. This was due in part to the lack of infrastructure, which made overland transport complicated and time consuming and to unreliable telecommunications (Salisbury, 1986). The hydroelectric project marked the beginning of cooperation among the communities of Eeyou Istchee and tight relations that were further facilitated by the intensive transport and communication infrastructure developments that followed. This period represents a sharp shift in Cree society, with changes to diet, physical activity, lifestyle and economy, not the least of which included the sedentarisation of the population in fixed villages (La Rusic, 1979; Desbiens, 2004). However, the large-scale assimilation predicted to be the outcome of these changes by many researchers did not materialise. Instead, like the phenomenon seen in many other northern indigenous communities in Canada, the Cree of Eeyou Istchee developed a unique culture through the contact of two interrelated spheres: subsistence activity and the market sphere, each with its own activities, institutions and traditions (Tanner, 1979; Feit, 1982; Bender and Morris, 1991; Hornig, 1999; Usher, 2003; Petit, 2010).

The Cree identity is heavily linked to the physical environment of Eeyou Istchee, as many social practices, traditions and values are based around their ancestral lands. The protection and transmission of this heritage is thought to be the responsibility of every member of the community; by extension, so is the protection of the territory (CTACC, 1989; CTA, 2009; Otis and Motard, 2009). The Cree's traditional subsistence activities include trapping, hunting, fishing and gathering and are interrelated with Cree values, cultural memory, knowledge and spirituality; they thus extend beyond simple economic activities to become integrated fully into Cree identity

(Scott and Feit, 1992; Tanner, 2007). To this day, a third of the Cree population of Eeyou Istchee has a lifestyle based on subsistence activities, and many more divide their time between traditional and market-based activities (Delormier and Kuhnlein, 1999).

The Cree territory of Eeyou Istchee is subdivided into family traplines that were officially recognised in the JBNQA (Morantz, 1978; Tanner, 1983; Fraser et al., 2006). One member of each family is selected per trapline to act as tallyman (or *Kaanoowapmaakin*). His role is to ensure the trapline's good management, the conservation and equitable sharing of the resources, the continuation through the generations of the trapline's practices and memories through stories and knowledge of the environment. Often selected because of his knowledge of Cree traditions and rules, his knowledge of the trapline's environment and ecosystem and his competence at traditional subsistence activities, a tallyman's role is strongly linked to diplomacy, and the rules are not always enforced in an overt manner (Scott, 1986; Feit, 1991; Berkes, 1998). The tallymen are helped in this by oral traditions that in 2009 were put into writing by the Cree Trappers' Association (CTA) to ensure their conservation. These oral traditions are passed down from one generation to the next, and in this context Elders are seen as playing an important role due to their knowledge of Cree history and traditions and are thus consulted in case of conflicts or disagreement between members of the community (CTA, 2009).

This integration between culture and environment as well as the long-term use of the Eastern James Bay by the Cree have allowed for the development of a detailed and profound knowledge of the local environment, climate and weather and the region's fauna and flora. This is particularly true in the case of tallymen and those that actively practice subsistence activities. Through continuous use and detailed knowledge of the area, they are more likely to detect changes and anomalies occurring due to changes in the local environment or climate.

The Eastern James Bay and the context of climate and environmental change

Any research on the impact of climate change must first take into account the context against which these changes have taken place. In the Eastern James Bay region, the development of extensive infrastructure has had serious implications for flora and fauna distribution and habitat in the area. In this section, these developments are reviewed as a means of scene setting for better disentangling and understanding the distinctive effects of climate changes in the region.

During the last ice age, the Laurentide Ice Sheet covered the entirety of this region. When it retreated approximately 8,000 years ago, it left the Hudson Bay's ancestor, the Tyrrell Sea. Both the Laurentide Ice Sheet and the Tyrrell Sea are responsible for many of the features now seen in the area;

the ice sheet being responsible for the territory's many lakes, while the re-
treat of the Sea left behind large marine clay deposits (Bider, 1976; Richard,
1979). Today the Eastern James Bay's climate is known for its extreme seasonal temperature variations, which can reach up to +34°C during the summers and drop down to −50°C in the winter (Knight, 1968; Laverdière and
Guimont, 1981). As per the Koeppen Climate Classification System, the
region's climate is moist continental mid-latitude. The James and Hudson
Bays to the west have a stabilising influence on precipitation keeping it to an
average of 765 mm/year (MBJ, 2010).

The western part of the James Bay Territory comprises large flatlands
dotted with peatlands (Richard, 1979; Thibault and Payette, 2009). The territory's central plateau is spotted with many lakes while the terrain becomes
more mountainous further to the east. A Boreal forest composed mainly of
black spruce (*Picea mariana*) and tarmarack larch (*Larix laricina*) covers
most of the area (Hare, 1950; Berkes and Farkas, 1978; Arquilière et al.,
1990). Further to the north, the dense boreal forest is replaced by the wooded
tundra with its vast spans of lichen dotted with diminutive trees that rarely
exceed 15 cm in diameter (Payette et al., 2001; Callaghan et al., 2002). The
Eastern James Bay is the habitat of 39 animal species, many of which such,
as the beaver (*Castor canadensis*) and the caribou (*Rangifer tarandus*), are
used as traditional food sources by the Cree. However, environmental factors including the region's climate are limiting factors to tree cover and animal populations.

Up until the 1960s, most of the Eastern James Bay had not undergone
many human-led changes. The southern part of the region had been developed for mining since the 1930s, but this activity was relatively limited in
scope. This changed with the JBNQA agreement and the start of a number
of development projects. The La Grande complex mega project is the best
known of these but only represents a downscaled version of the original
plan announced by the government in 1971. Started in 1973, the reworked La
Grande complex mega project was divided into three phases. The first two
phases included diverting the Caniapiscau, Eastmain and Opinaca rivers
to feed eight hydroelectric power plants and their 13,575 km^2 of reservoirs
(Hydro-Québec, 2003). The diversion of these three rivers modified the local
hydrological environment by flooding 9,900 km^2 of land and by significantly
reducing the three rivers' flow while increasing the winter flow of the La
Grande River (Reed et al., 1996; Hernandez-Henriquez et al., 2010). In the
case of the Eastmain River, the flow at the mouth of the river was perma-
nently reduced by 90 percent, which lowered water levels and increased the
river's salinity from a larger intake of salt water from the bay. Work on the
project's third phase, the Eastmain-1-A-Sarcelle-Rupert, started in January
2007. This phase consists of creating two new power stations, Eastmain-1A
and Sarcelle, and diverting 51.7 percent of the flow of the Rupert River
towards the new 600 km^2 Eastmain Reservoir and existing La Grande
Complex (Hydro-Québec, 2004).

This hydroelectric project also required the wholesale creation of extensive infrastructure and networks. The energy distribution towards the south is assured by six 735 kV overhead lines that crisscross the James Bay Territory from North to South (Hydro-Québec, 2010). The project also included the construction of the 620-km-long James Bay Road, the 666-km-long Transtaïga Road and the 407-km-long Route du Nord, as well as a number of smaller roads linking the Cree Communities and the new locality of Radisson to the three main roads. Finally, the project included the creation of a number of airports, buildings and telecommunication infrastructure. The creation of a roadway network across the territory also paved the way for intensive mining exploration in the area starting in the 1970s. Part of the Canadian Shield, the area was quickly found to contain important quantities of gold, copper, zinc and silver, while more recent explorations have also hinted at diamond, lithium and uranium deposits (Raymond, 2011; SDBJQ, 2011). The region's active mines are contained in the southern portion of the territory; however, current explorations have ventured further north. The southern part of the James Bay Territory also has a thriving lumber industry covering 81,158 km^2. It is constrained to the southern portion as the territory is on the limit of commercially available forests (MRNF, 2012). Against this backdrop of rapid and dramatic anthropogenic environmental modification, climate changes have taken place. The indicators, observations and implications of these changes are explored in the following section.

Observing environmental change

The Cree of the Eastern James Bay practice subsistence activities at different times of the year. Whereas fishing and gathering are practiced mainly during the warmer months, hunting and trapping are at their peak over the colder months. For the Cree located inland, the most intensive hunting occurs over the winter, while for coastal Cree the autumn and spring hunting seasons are as important. Cree trappers are, therefore, aware of the changing seasons and weather conditions as they wait for the appropriate time of year and environmental conditions to practice their activities. There are six broad categories of hunted animals as defined by the Cree. These are large mammals, which include the caribou and bear, fish, geese such as the Canada geese and the white geese; sea mammals, including the polar bear and seal; small mammals, such as the porcupine and hare; and finally fur mammals, like the beaver or muskrat (Berkes and Farkas, 1978). Although the proportion of each animal in a typical diet can vary from year to year depending on a multitude of factors, geese represent for coastal communities one of the most important sources of wild dietary meat. This hunt occurs mainly during their migration across the territory in the autumn and spring seasons. The northernmost communities and traplines rely on caribou that descend south from the Arctic over the winter while the southern communities count more heavily on the beaver (Berkes, 1990). During hunting season, and particularly during the goose-hunting period, locally called 'Goose Break', the tallymen often become the hunt supervisors. They are tasked with insuring the diffusion of the hunting pressure in time and space. They select hunting locations based on a number of elements including human factors, such as the number of hunters and their capabilities, biological factors like the behaviour of the animals and the location of their food sources, and climatic conditions, such as temperature, wind strength and direction. Movement across the land for hunting is often undertaken using the multiple rivers and lakes that crisscross the area after they have frozen over during the colder months. Because of this hunting pattern, it would be safe to assume that Cree hunters are more intensively attuned to weather conditions occurring from autumn to spring.

Information presented here was collected through interviews and questionnaires carried out in 2010 with adult members of the Cree Trappers Association, a Cree organisation whose goals include "to foster, promote, protect and assist in preserving the way of life, values, activities and traditions of the Cree trappers of Quebec and to safeguard the system of the Cree traplines" (CTA, 2012). Participants for the short interviews ($n = 18$) and the questionnaires ($n = 30$) were selected using a non-probabilistic accidental sampling method, while in-depth interview participants ($n = 3$) were key informants identified using a non-probabilistic snowball sampling method.

We undertook our research with this organisation's members because they still practice traditional subsistence activities and are interested in their conservation. It is telling that 73.5 percent of the adult Cree population were members in 2010 (CTA, 2010a). We asked participants direct questions about changing weather conditions as well as indirect questions about changing animal habitats and behaviour. Participants were also asked questions about modifications to their hunting habits and the transmission of the practice of traditional subsistence activities to younger generations.

Participants identified changes using a variety of instruments and tools such as temperature and ice thickness, as well as clues in the physical environment garnered from observations of, for example, the direction and speed of incoming clouds to identify possible storms. Elder participants also mentioned observations based on animal behaviour: "One bird, he keeps on making noise depending on the weather. The Cree people understand what the birds are saying. He is telling the people what the weather is going to be" (Anonymous, 2010). This informant explained that he started hunting at age 12, which is also the time he started his training as tallyman. He described doing the same with his children. It is telling that in key places Cree hunter observations based on TEK and scientific knowledge coalesced on certain environmental circumstances. This is, for example, the case when defining ice conditions. The Cree distinguish two types of soft water ice: black ice and white ice, which correspond in the scientific literature to congelation ice and snow ice. They also pay attention to water pockets and pools of water on the surface of the ice, which are associated with the softer white/snow ice and ice shards/candles/vertical ice structures with the stronger black/congelation ice. The distinction between these two types of ice is important as they have different load-bearing strengths.

Throughout our questions about changing weather questions, although many mentioned warming weather and later winters, one of the most significant observations focused on an increased quantity of rain during the colder months. This can be mixed with snowfalls up until December and then start up again in the spring as early as March. As participants were asked to compare the current situation with a fixed point in time, they often gave a general impression of the month. This means that a month that in the past was cold and snowy and today had an increased number of rainy days would therefore seem warmer, although this might not be completely reflected in temperature data (Royer, 2016). We also received detailed answers concerning ice conditions on the rivers and lakes. Participants mentioned that the ice characteristics had changed, becoming weaker. They used descriptions such as there being more and more 'soft ice'; "there are no more icicles in the ice" (Anonymous, 2010), "there is less ice" (Anonymous, 2010) or "The ice shards are gone and there are more water pockets and pools in the ice" (Anonymous, 2010). Ice shards or ice candles are vertical structures in the ice. Although an intrinsic part of the structure of congelation ice, they can often best be seen in the spring when the ice begins to thaw. By contrast, water pockets and a white coloured ice are signs of white ice or snow ice. By combining this information with instrumental weather data, the results seemed to point to the increased precipitation in autumn, winter and spring as a possible cause for this increased presence of snow ice in the Eastern James Bay (Royer et al., 2013). Travel over ice is often most risky at its highest in the autumn (freeze-up) and spring (break-up) periods as the ice hasn't fully formed or has already begun to thaw and, therefore, has a lower load-bearing capacity.

This interest in ice conditions could be linked to an increased awareness of hazards linked to changing ice conditions due to changing weather conditions, as was stated in 2010 in the CTA Whapmagoostui Community Report: "In spring we lost two very experienced hunters that fell through the ice" and again in 2011 by the CTA vice-president: "Awareness programming is very important because in the last five years Mistissini alone has had three serious accidents on the ice, all involving experience hunters. [...] So we have to advertise that people must be very careful and exercise extra caution, especially in late spring" (CTA, 2010b; McDonagh, 2011). Injury and heart disease have long been identified as the most common causes of death for the Eastern James Bay Cree (Robinson, 1988). For such small and close-knit communities as the ones in the Eastern James Bay, these losses are felt profoundly, with the whole community coming together to mourn.

Serious accidents are always worrisome but even more so when those accidents involve experienced hunters who through their familiarity with the environment and local weather conditions should be more aware of identifying potential hazards. From the interviews conducted for this research, along with other similar interviews with hunters, it was noted that extreme weather events such as storms and heavy rain have become harder to predict. One participant admitted that he had recently been nearly caught by a storm while out hunting because all his normal weather markers were pointing to calm weather for that day. He explained that he noticed the storm with just enough time to take shelter and was worried about what would have happened had he not been as quick to act. Similar information was reported during a community workshop held by the CTA in Whapmagoostui with members stating that "It is more difficult to predict the weather now" (CTA, 2010b). Members commenting on this change often pointed to two contributing factors. The first factor often concerned younger hunters and was that they were reliant on technology when they were out on the land and were not paying as much attention to smaller observational details. The second was that there was a change in the weather patterns, and thus traditional markers were sometimes misleading because the weather was not behaving in the usual way.

Participants also made wider observations about the environment. One participant, when told about our research subject, instinctively confided: "You know what I've noticed about climate change... there are less frogs" (Anonymous, 2010). Data from the Cree geoportal (a community-led, GIS-based tool created by the CTA for reporting observations of environmental change in the Eastern James Bay) and regional coastal survey data support this observation of a decline in frog species (Desroches et al., 2011; Herrmann et al. 2012). In Eastmain, part of the decrease in frog population can be traced back to the diversion of the Eastmain River and its increased salinity. However, similar decreases were noted in inland and southern areas of the James Bay Territory. Although the frog as a species is not part of Cree subsistence activities, it can serve as a valuable indicator of climate and environmental change, given that frogs are attuned to more minute changes in the environmental and climatic conditions (Wake and Vredenburg, 2008).

Changes in the practice of subsistence activities

Questions about the practice of subsistence activities asked Cree hunters to identify changes occurring through the behaviour modifications of two hunted species, the Canada goose and the woodland caribou. They were also invited to compare their current habits to those they had as young adults. Finally, they were questioned on changes between their generation and the ones following. As discussed earlier, the region has undergone dramatic changes to its social and physical environments, and it is important to keep those in perspective when looking at changes in animal and human behaviour.

The Canada goose was looked at in particular because it is an important staple of the traditional Cree diet. Geese migration patterns have been studied as indicators of climate change (Ball, 1983). All participants ate Canada goose meat regularly, and all but two reported hunting it. However, as Cree hunters are not required to report geese kill numbers, the only way to know if there is a decrease or increase in the number of kills is through self-reporting. As a guide, a report from 1975 pointed to a ratio of 10.5 Canada geese per person (Scott and Feit, 1992). Questions on consumption of goose meat indicated a decrease only for participants aged 40 to 59, and this trend was highest for coastal participants. As people in this age group reported being more active in the market economy and having less time to hunt, it is safe to assume that this decrease is related to social rather than environmental changes (Royer and Herrmann, 2011). Cree hunters' responses also pointed to an inland movement of the Canada geese. This could be deduced from coastal respondents observing and killing a decreased number of geese while inland respondents mentioned observing and killing an increased number. Participants identified the causes for these changes being linked to a change of flight path: "because the flight path is different; more and more the flight paths change" (Anonymous, 2010). The geese also prefer the seagrass-rich marshlands to large open waterscapes: "They won't stay in areas that are now open water" (Anonymous, 2010). In this case again, therefore, it would be difficult to blame weather conditions without further and more in-depth study. The final key observation about the Canada geese in the territory concerned subspecies identification. Current government-led studies in the region only identify one type of subspecies in the region: the interior Canada goose called short necks by the Cree (*Branta canadensis interior*). However, almost half of respondents mentioned having seen or killed a subspecies usually found further south: the resident Canada goose called long necks by the Cree (*Branta canadensis maxima*). The migration of long necked geese to the north (with some hunters noting that the species had been observed as high as the Hudson Bay) is seen as an undesirable effect of climate change by the hunters as the long necks are thought to be less tasty than the short necked geese. Studies conducted on Canada geese found that changes in climatic and environmental conditions could also have an effect on the geese morphology, thus an increased presence of food could cause the better-fed short necked geese to resemble their long necked siblings (Leafloor et al., 1998).

Because of the trapline system in the Eastern James Bay Territory, long-term changes in the distribution of geese can cause modifications to the practice of subsistence activities as families that traditionally had access to geese on their lines are confronted with decreasing numbers. These families would then have to ask permission to hunt on other traplines, thus having an impact on social dynamics inside the community. The potential decrease in hunting opportunities for some families linked with an undesirable change in subspecies or the geese's taste of goose can bring about important impacts on the Cree diet if it becomes associated with a decrease in nutrition-rich geese consumption.

Woodland caribou in the north of the James Bay Territory are identified as a migratory species. Further south, in the boreal forest, the same species is sedentary and is considered an 'at risk' species due to their continued decline in line with their habitat degradation (COSEWIC, 2002). Before the population evaluation of 2010, the migratory species, in contrast to its sedentary cousin was considered a very healthy population. However, alarm bells followed the 2010 evaluation, which showed an estimated 80-percent decline in the population from a high of 1,013,000 animals in 2001 (Courtois et al., 2003; Boulet et al., 2005; MRNF, 2010, 2011). With reports of the decrease continuing, the situation is worrisome. This decrease in caribou populations was also noted in Cree participants' observations, with half indicating decreased numbers. Some members mentioned that the caribou populations followed a cyclical trend: "You see them, then you have to wait seven years to see them again" (Anonymous, 2010). While this cyclical behaviour is known, and although it is longer than seven years, it has never been shown to cause such a decline in numbers.

The causes for this sharp decrease in caribou populations have not been identified. However, it has been theorised that it could be at least partially linked to changes in weather patterns. For example, the potential effects of weather and climate change in the area can include vegetation modifications leading to a decrease in habitat and food, an increase in parasites, greater competition for resources from an increased presence of other species and an increase in predators. Certain participants also noted having observed increased caribou numbers travelling further south, mentioning that the caribou now like to follow the new roads and clear-cut areas below the hydro-electric transmission lines: "[...] they [caribou] are not eating the right kind of food up North. They are eating under the powerlines and their food is all underwater, that is why they are now moving around here to Nemaska" (Anonymous, 2010). They also mentioned that the animal's fur was now patchy and that the meat didn't taste the same; that is was less good because the animal was eating different food sources.

With climate change being theorised to bring about increased snow and rainfall as well as heightened risk of forest fires, the potential of further damage to caribou populations through food limitations is high (Tyler, 2010). The animals were also reported to be thinner by the Cree participants, which studies have linked to a longer presence of biting insects due to longer warmer weather seasons (Toupin et al., 1996; CTA, 2010b). When questioned on their caribou hunting habits, as expected inland and northern members were the ones that most hunted caribou. However, the majority of participants mentioned a decrease in the number of kills. Reported numbers of caribou kills correspond to the self-reported data, with reported numbers for the whole of Eeyou Istchee going from one caribou per seven members in the 2008–2009 hunting season to one per 37 members for the 2009–2010 hunting season (Royer and Herrmann, 2011). Like the Canada geese, changes in caribou hunting can have an effect on the practice of subsistence activities and thus in turn on social dynamics as well as on the health and nutrition of the Cree that depend on it. This noticeable decrease in caribou hunting could represent in time a loss to the associated cultural memories and knowledge.

Although as within any society many informal comments can be heard about younger generations not being as active as older ones, half of the participants responded that the younger generations hunted geese as much as their own. Of those that mentioned they hunted less, some explained that it was due to the decreased number in their area. The closer the participant's generation was to the younger one, the less likely he was to note a difference in hunting/eating habits and vice versa. This trend didn't hold true for the caribou, as the majority noted that younger generations hunted and consumed this animal species either to the same extent as previous generations or less. As TEK and knowledge in subsistence activities are often transmitted through the practice of the activity, the decrease shown among the older generations would theoretically translate into a decrease in all generations following it (Ohmagari and Berkes, 1997; Parlee and Berkes, 2006). Increased participation with the southern economy has also had an impact on the transmission of the production of many hunting-associated goods, with a participant explaining "No one is doing that anymore [goose down pillows and blankets], all because you can buy these items from stores nowadays" (Anonymous, 2010). As for any traditional craft in most societies, this trend is to be expected. By comparing habits linked to two species, one decreasing in numbers and the other maintaining its numbers, results seem to point to decreases in practices being linked to decreased availability of the species, especially when looking at key dietary species.

Conclusions

Cree TEK, and TEK in general, expands beyond the limits of simply knowing. It is intrinsically linked to cultural memories of time and place. Traditional knowledge and the practice of traditional activities are part of the Cree cultural heritage and as such are linked to their collective identity. The importance of the continued practice of subsistence activities in relation to mental and physical health has been shown in a number of Aboriginal communities. As the practice of these activities is linked to general health and well-being, the decrease in this practice has been linked to an increase in negative social problems such as suicide (Parlee et al., 2005; Tousignant et al., 2008). Cree TEK's link with space is undeniable. This can in part be seen through the tallyman's role in remembering and incorporating the history and environment of the family's traplines into the practice of subsistence activity. The role of tallyman and his management of family traplines is to this day a central part of traditional Cree culture. He is tasked with knowing these *lieux de mémoire* and transferring this knowledge to the younger generations.

As these activities are linked very closely with the environment and the weather, they have also come to play an important role in cultural memory. Memories are linked to the weather conditions present at the time of the practice of activities. The start of the autumn and spring goose hunts must wait until favourable weather conditions allow for the correct combination of factors in the bush. The change in weather conditions that signals the start of the 'goose break' is awaited eagerly by many and is linked to excitement "similar to Christmas in the south" (Anonymous, 2010). In addition, the environment's fauna and flora also offer clues to understanding weather patterns and, as we have seen, this understanding of the clues offered in nature is seen as part of Cree cultural identity.

The changing climate and with it the changing weather represent real and important challenges for the Cree. The increased hazards linked to the practice of subsistence activities through the modification of environmental conditions require the Cree to quickly integrate new behaviours and security measures while animal presence and behaviour modifications could lead to changes in community dynamics and changes in the key species for subsistence activities. For the Cree of the Eastern James Bay Territory, the impacts of climate change over the last 100 years are superimposed onto a period of intense socio-environmental and economic change. Through all of this, subsistence activities have continued to be practiced regularly by community members, with the Cree incorporating and adapting, as they have done before, new knowledge and technologies into their TEK.

References

Anonymous (2010) In-depth and short interview CTA hunters, Québec. Interviewed by: Marie-Jeanne Royer, August 2010.

Arquilière S, Filion L, Gajewski K and Cloutier C (1990) A dendroecological analysis of eastern larch (*Larix laricina*) in subarctic Quebec. *Canadian Journal of Forest Research*, 20: 1312–1319.

Ball T (1983) The migration of geese as an indicator of climate change in the southern Hudson Bay region between 1715 and 1851. *Climatic Change*, 5: 85–93.

Bender B and Morris B (1991) Twenty years of history, evolution and social change in gatherer-hunter studies. In Ingold T, Riches D and Woodburn J (eds.) *Hunters and Gatherers, Volume 1: History, Evolution and Social Change*. Oxford: Berg Publishers Limited: 4–14.

Berkes F (1990) Native subsistence fisheries: a synthesis of harvest studies in Canada. *Arctic*, 43 (1): 35–42.

Berkes F (1998) Indigenous knowledge and resource management systems in the Canadian subarctic. In Berkes F and Folke C (eds.) *Linking Social and Ecological Systems: Management Practices and Social Mechanisms for Building Resilience*. Cambridge: Cambridge University Press: 98–128.

Berkes F and Farkas CS (1978) Eastern James Bay Cree Indians: changing patterns of wild food use and nutrition. *Ecology of Food and Nutrition*, 7 (3): 155–172.

Bider JR (1976) The distribution and abundance of terrestrial vertebrates of the James and Hudson Bay Regions of Québec. *Cahiers de géographie du Québec*, 20 (50): 393–407.

Boulet M, Couturier S, Côté SD, Otto R and Bernatchez L (2005) Flux génique entre les troupeaux de caribous migrateurs, montagnards et sédentaires du Nord-du-Québec et du Labrador: repérages par satellite, génotypage de microsatellites et simulation de populations. Ministère des Ressources naturelles et de la faune, Direction de la recherche sur la faune, Québec, Canada.

Callaghan TV, Crawford Rober MM, Eronen M, Hofgaard A, Payette S, Rees G W, Skre Oddvar, Sveibjörnsson B, Vlassova TK and Werkman BR (2002) The dynamics of the Tundra-Taiga boundary: an overview and suggested coordinated and integrated approach to research. *Ambio*, Special Report 12: 3–5.

Canobbio É (2009) *Géopolitique d'une ambition Inuite. Le Québec face à son destin nordique*. Quebec: Septentrion.

Catchpole AJW (1992) Hudson's Bay Company ships' log-books as sources of sea ice data, 1751–1870. In Bradley Raymond S and Jones Philip D (eds) *Climate Since AD 1500*. London and New York: Routledge: 17–39.

CCRCCN – comité chargé du réexamen de la commission Crie-Naskapie (1991) Report of the inquiry into the Cree-Naskapi Commission, Canada.

Chaplier M (2006) Le conflit à la baie James: Pour une anthropologie de la nature dans un contexte dynamique. *Civilisations*, 55: 103–115.

COSEWIC (2002) *Assessment and Update Status Report on the Woodland Caribou Rangifer tarandus caribou in Canada*. Ottawa: Committee on the Status of Endangered Wildlife in Canada.

Courtois R, Ouellet J-P, Gingras A, Dussault C, Breton L and Maltais J (2003) Historical changes and current distribution of caribou, Rangifer tarandus, in Quebec. *Canadian Field-Naturalist*, 117 (3): 399–414.

CRBJ – Conférence régionale de la Baie-James (2011) Portrait de la Jamésie, 2011. Quebec. www.crebj.ca (accessed 7 June 2011).

CTA – Cree Trappers Association (2009) Traditional Eeyou Hunting Law. Eastmain Canada, unpublished.

CTA – Cree Trappers Association (2010a) Activity Report 2009–2010. Eastmain Canada, unpublished.

CTA – Cree Trappers Association (2010b) The climate change project: impact and adaptation for the hunters, trappers and communities of Eeyou Istchee, Whapmagoostui Community Report – April 2010. Eastmain Canada, unpublished.

CTA – Cree Trappers Association (consulted 2012) Objects of the Cree Trappers' Association, The Incorporation Papers of the Association recorded March 31, 1978. www.creetrappers.ca/objects.php (accessed 12 June 2012).

CTACC – Cree Trappers Association's Committee of Chisasibi (1989) Cree trappers speak. James Bay Cree Cultural Education Center, Quebec, Canada.

Delormier T and Kuhnlein HV (1999) Dietary characteristics of Eastern James Bay Cree Women. *Arctic*, 52 (2): 182–187.

Desbiens C (2004) Producing North and South: a political geography of hydro development in Québec. *The Canadian Geographer*, 48 (2): 101–118.

Desroches F, Schueler FW, Picard I and Gagnon LF (2011) A herpetological survey of the James Bay area of Québec and Ontario. *The Canadian Field Naturalist*, 124 (4): 299–315.

Ducruc J-P, Zarnovican R, Gerardin V and Jurdant M (1976) Les régions écologiques du territoire de la baie de James: caractéristiques dominantes de leur couvert végétal. *Cahiers de Géographie du Québec*, 20 (50): 365–392.

Duhaime G (ed.) (2001) *Le Nord: Habitants et Mutations*. Quebec: Les Presses de l'Université Laval.

Feit HA (1982) The future of hunters within Nation-States: anthropology and the James Bay Cree. In Leacock EB and Lee R (eds.) *Politics and History in Band Societies*. Cambridge: Cambridge University Press: 373–412.

Feit HA (1991) Gifts of the land: hunting territories, guaranteed incomes and the construction of social relations in James Bay Cree Society. *Senri Ethnological Studies*, 30: 223–268.

Feit HA (1995) Hunting and the quest for power: the James Bay Cree and Whitemen in the 20th century. In Morrison BR and Wilson RC (eds.) *Native Peoples: The Canadian Experience*, 2nd edition. Toronto, ON: McClelland & Stewart Publishers: 101–128.

Fraser DJ, Coon T, Prince MR, Dion R and Bernatchez L (2006) Integrating traditional and evolutionary knowledge in biodiversity conservation: a population level case study. *Ecology and Society*, 11(2): art4.

Froschauer K (1999) *White Gold: Hydroelectric Power in Canada*. Vancouver, BC: UBC Press.

Goudreau É (2003) Les autochtone et le Québec. In Weidmann-Koop M-C (ed.) *Le Québec aujourd'hui: identité, société et culture*. Quebec: Les Presses de l'Université Laval: 121–140.

Hare KF (1950) Climate and zonal divisions of the Boreal forest formation in Eastern Canada. *The Geographical Review*, 40 (4): 615–635.

Hernandez-Henriquez MA, Mlynowski TJ and Dery SJ (2010) Reconstructing the natural streamflow of a regulated river: a case study of La Grande Riviere, Quebec, Canada. *Canadian Water Resources Journal*, 35 (3): 301–316.

Herrmann TM, Royer MJS and Cuciurean R (2012) Understanding subarctic wildlife in Eastern James Bay under changing climatic and socio-environmental conditions: bringing together Cree hunters' ecological knowledge and scientific observations. *Polar Geography*, 35 (3–4).

Hogue C, Bolduc A and Larouche D (1979) *Québec, un siècle d'électricité*. Montreal, QC: Libre Expression.

Hornig JF (ed.) (1999) *Social and Environmental Impacts of the James Bay Hydroelectric Project*. Montreal, QC: McGill-Queen's University Press.

Hydro-Québec (2003) *La Grande Hydroelectric Complex: fish communities*. Quebec, Canada.

Hydro-Québec (2004) *Centrale de l'Eastmain-1-A et dérivation Rupert: Étude d'impact sur l'environnement – Rapport de synthèse*. Québec, Canada.

Hydro-Québec (2010) *Rapport Annuel 2010: grands équipements*. Québec, Canda.

Knight R (March 1968) Ecological factors in changing economy and social organization among the Rupert House Cree. Anthropology Papers, National Museum of Canada, Department of the Secretary of State, Ottawa, Canada.

150 *Marie-Jeanne S. Royer*

La Rusic IE (ed.) (1979) *La négociation d'un mode de vie: la structure administrative découlant de la Convention de la Baie James: L'expérience initiale de Cris.* Montreal, QC: ssDcc Inc.

Laverdière C and Guimont P (1981) Géographie physique de la Grande île, Littoral québécois de la Mer d 'Hudson. Société de développement de la Baie James, Aménagement régional, Québec, Canada.

Leafloor JO, Davidson AC and Rusch DH (1998) Environmental effects on body size of Canada geese. *The Auk*, 115 (1): 26–33.

MAMROT – Ministère des Affaires municipales régions et occupation du territoire du Québec (2011) Baie-James, 2011. Quebec. www.mamrot.gouv.qc.ca (accessed 6 June 2011).

MBJ – Municipalité de Baie-James (2010) Territoire de la Baie-James, 2010. Quebec. www.municipalite.baie-james.qc.ca/html/territoire_bj.php (accessed 25 July 2011).

McDonagh P (2011) Observing the effects of climate change in Eeyou Istchee. Public Health Department o the Cree Health Board, Quebec, Canada. www.creehealth. org/news/observing-effects-climate-change-eeyou-istchee (accessed 25 November 2011).

MDDELCC – Ministère du Développement durable, Environnement et Lutte contre les changements climatiques (2016) Région administrative du Nord-du-Québec, Portrait socio-économique de la région, 2002. Quebec. www.mddelcc.gouv.qc.ca/regions/region_10/portrait.htm (accessed 19 October 2016).

Morantz T (1978) The probability of family hunting territories in eighteenth century James Bay: old evidence newly presented. In Cowan W (ed.) *Proceedings of the Ninth Algonquian Conference*. Ottawa: Carleton University.

Morantz T (2002) *The White Man's Gonna Getcha: the Colonial Challenge to the Crees in Quebec*. Montreal, QC and Kingston, ON: McGill-Queen's University Press.

MRNF – Ministère des Ressources naturelles et de la Faune du Québec (9 November 2010) Résultat de l'inventaire du troupeau de caribous de la rivière George – Le ministre Simard préoccupé par la réduction du Cheptel. Québec. www.mrnf. gouv.qc.ca/presse/communiques-detail.jsp?id=8714 (accessed 31 May 2011).

MRNF – Ministère des Ressources naturelles et de la Faune du Québec (11 November 2011) Résultat de l'inventaire du troupeau de caribous de la rivière aux Feuilles – Le ministre Simard suit de près la décroissance du cheptel. Québec. www.mrnf. gouv.qc.ca/presse/communiques-detail.jsp?id=9382 (accessed 31 December 2011).

MRNF – Ministère des Ressources naturelles et de la Faune du Québec (consulted 2012) Limite nordique des forêts attribuables pour un aménagement forestier durable. Québec. www.mrnf.gouv.qc.ca/mines/quebec-mines/2011-06/lithium.asp (accessed 12 June 2012).

Niezen R (1993) Power and dignity: the social consequences of hydro-electric development for the James Bay Cree. *Canadian Review of Sociology and Anthropology*, 30: 510–529.

Ohmagari K and Berkes F (1997) Transmission of Indigenous knowledge and bush skills among the western James Bay Cree women of subarctic Canada. *Human Ecology*, 25 (2): 197–222.

Otis G and Motard G (2009) De Westphalie à Waswanipi: la personnalité des lois dans la nouvelle gouvernance crie. *Les Cahiers de Droit*, 50 (1): 121–152.

Parlee B and Berkes F (2006) Indigenous knowledge of ecological variability and commons management: a case study on berry harvesting from Northern Canada. *Human Ecology*, 34: 515–528.

Parlee B, Berkes F and Teetl'it Gwich'in Renewable Resources Council (2005) Health of the land, health of the people: a case study on Gwinch'in berry harvesting in Northern Canada. *EcoHealth*, 2: 127–137.

Payette S, Fortin M-J and Gamache I (2001) The subarctic forest-tundra: the structure of a biome in a changing climate. *BioScience*, 51 (9): 709–718.

Peloquin C and Berkes F (2009) Local knowledge, subsistence harvests and social-ecological complexity in James Bay. *Human Ecology*, 37: 533–545.

Petit J-G (2010) Cris et Inuit du nord du Québec: Deux peoples entre tradition et modernité (1975–2010). In Petit J-G, Bonnier YB, Aatami P and Iserhoff A (eds.) *Les Inuits et les Cris du Nord du Québec*. Rennes: Presses Universitaires de Rennes: 15–27.

Raymond D (2011) Le lithium au Québec: les projets miniers d'actualité, Ministère des Ressources naturelles et de la Faune, Québec: juin 2011. Gouvernement du Québec, Québec. www.mrnf.gouv.qc.ca/mines/quebec-mines/2011-06/lithium.asp (accessed 15 July 2011).

Reed A, Benoit R, Julien M and Lalumière R (1996) *Utilisation des habitats côtiers du nord-est de la baie James par les bernaches*. Publication hors-série No92, Service Canadien de la Faune, Canada.

Richard P (1979) Contribution à l'histoire postglaciaire de la végétation au nord-est de la Jamésie, Nouveau Québec. *Géographie Physique et Quartenaire*, XXXIII (1): 93–112.

Richardson B (1976) *Strangers Devour the Land*. New York: Alfred A Knopf.

Robinson E (1988) The health of the James Bay Cree. *Canadian Family Physician*, 34: 1606–1613.

Royer MJS (2016) *Climate, Environment and Cree observations: James Bay Territory, Canada*, SpringerBriefs in Climate Studies, Switzerland: Springer International Publishing.

Royer MJS and Herrmann TM (2011) Socio-environmental changes in two traditional food species of the Cree First Nation of subarctic James Bay. *Cahiers de géographie du Québec*, 55 (156): 575–601.

Royer MJS, Herrmann TM, Sonnentag O, Delusca K, Fortier D and Cuciurean R (2013) Linking Cree Hunters' and scientific observations of changing inland ice and meteorological conditions in the subarctic Eastern James Bay region, Canada. *Climatic Change*, 119 (3): 719–732.

SAA – Secrétariat aux Affaires autochtones (1998) *James Bay and Northern Québec Agreement and Complementary Agreements*, 1998th edition. Québec: Les Publications du Québec.

SAA – Secrétariat aux affaires autochtones (2009) Profil des Nations: Cris, Québec: 19 mai 2009. Quebec. www.saa.gouv.qc.ca/realtions_autochtones/profils_nations/cris.htm (accessed 4 August 2011).

Salisbury RF (1986) *A Homeland for the Cree: Regional Development in James Bay, 1971–1981*. Montreal, QC: McGill-Queen's University Press.

Savard S (2009) Le communautés autochtones du Québec et le développement hydroélectrique: un rapport de force avec l'État, de 1944 à aujourd'hui. *Recherches Amérindiennes au Québec*, 39 (1–2): 47–60.

Scott C (1986) Hunting territories, hunting bosses and communal production among coastal James Bay Cree. *Anthropologica*, 28 (1/2): 163–173.

Scott C (1988) Property, practice and aboriginal rights among Quebec Cree hunters. In Ingold T, Riches D and Voodburn J (eds.) *Hunters and Gatherers, Volume 2: Property, Power and Ideology*. Oxford: Berg Publishers Limited: 35–51.

Scott C (1989) Ideology of reciprocity between the James Bay Cree and the Whiteman State. In Skalník P (Ed.) *Outwitting the State, Political Anthropology, Volume 7.* Piscataway, NJ: Transaction Publishers: pp. 81–108.

Scott C and Feit HA (1992) *Income Security for Cree Hunters: Ecological, Social and Economic Effects*, Monograph series. McGill Program in the Anthropology of Development, Montreal, QC, Canada.

SDBJQ – Société de développement de la Baie-James du Québec (consulted 2011) Les projets régionaux: projets miniers, Gouvernement du Québec, Québec: 2008–2009. Quebec. www.sdbj.gouv.qc.ca/fr/projets_developpement/projets_miniers (accessed 04 August 2011).

Statistics Canada (2012a) Oujé-Bougoumou, Québec (Code 2499818) and Nord-du-Québec, Quebec (Code 2499) (table). Census profile, 2011 census. Statistics Canada Catalogue no. 98-316-XWE. Ottawa, Canada.

Statistics Canada (2012b) Eastmain, Québec (Code 2499810) and Nord-du-Québec, Quebec (Code 2499) (table). Census profile, 2011 census. Statistics Canada Catalogue no. 98-316-XWE. Ottawa, Canada.

Statistics Canada (2012c) Chisasibi, Québec (Code 2499814) and Nord-du-Québec, Quebec (Code 2499) (table). Census profile, 2011 census. Statistics Canada Catalogue no. 98-316-XWE. Ottawa, Canada.

Statistics Canada (2012d) Mistissini, Québec (Code 2499804) and Nord-du-Québec, Quebec (Code 2499) (table). Census profile, 2011 census. Statistics Canada Catalogue no. 98-316-XWE. Ottawa, Canada.

Statistics Canada (2012e) Nemaska, Québec (Code 2499808) and Nord-du-Québec, Quebec (Code 2499) (table). Census profile, 2011 census. Statistics Canada Catalogue no. 98-316-XWE. Ottawa, Canada.

Statistics Canada (2012f) Waskaganish, Québec (Code 2499806) and Nord-du-Québec, Quebec (Code 2499) (table). Census profile, 2011 census. Statistics Canada Catalogue no. 98-316-XWE. Ottawa, Canada.

Statistics Canada (2012g) Waswanipi, Québec (Code 2499802) and Nord-du-Québec, Quebec (Code 2499) (table). Census profile, 2011 census. Statistics Canada Catalogue no. 98-316-XWE. Ottawa, Canada.

Statistics Canada (2012h) Wemindji, Québec (Code 2499812) and Nord-du-Québec, Quebec (Code 2499) (table). Census profile, 2011 census. Statistics Canada Catalogue no. 98-316-XWE. Ottawa, Canada.

Statistics Canada (2012i) Whapmagoostui, Québec (Code 2499816) and Nord-du-Québec, Quebec (Code 2499) (table). Census profile, 2011 census. Statistics Canada Catalogue no. 98-316-XWE. Ottawa, Canada.

Swyngedouw E (2007) Technonatural revolutions: the scalar politics of Franco's hydro-social dream for Spain, 1939–1975. *Transactions of the Institute of British Geographers*, 32 (1): 9–28.

Symon C, Arris L and Heal B (eds.) (2005) *Arctic Climate Impact Assessment.* Cambridge: Cambridge University Press.

Tanner A (1979) Bringing home animals: religious ideology and mode of production of the Mistassini Cree Hunter. Institute of Social and Economic Research, Memorial University of Newfoundland, Canada.

Tanner A (1983) Algonquin land tenure and state structures in the North. *Canadian Journal of Native Studies*, 3 (2): 311–320.

Tanner A (2007) The nature of Quebec Cree animist practices and beliefs. In Laugrand FB and Oosten JG (eds.) *La nature des esprits dans les cosmologies autochtones.* Québec: Les Presses de l'Université de Laval.

Thibault S and Payette S (2009) Recent permafrost degradation in bogs of the James Bay area, Northern Quebec, Canada. *Permafrost and Periglacial Processes*, 20: 383–389.

Toupin B, Huot J Manseau M (1996) Effect of insect harassment on the behaviour of the Rivière George Caribou. *Arctic*, 49 (4): 375–382.

Tousignant M, Laliberté A, Bibeau G and Noël D (2008) Comprendre et agir sur le suicide chez les Premières Nations: quelques lunes après l'initiation. *Frontières*, 21 (1): 113–119.

Trenberth KE, Jones PD, Ambeneje R, Bojariu R, Easternling D, Klein Tank A, Parker D, Rahimzadeh F, Renwick JA, Rusticci M, Soden B and Zhai P (2007) Observations: surface and atmospheric climate change. In Solomon S, Qin D, Manning M, Hen Z, Marquis M, Avery KB, Tognor M and Miller HL (eds.) *Climate Change 2007: The Physical Science Basis. Contribution of Working Group 1 to the Fourth Assessment Report of the Intergovernmental Panel on Climate Change.* Cambridge and New York: Cambridge University Press: 235–336.

Tyler NJC (2010) Climate, snow, ice, crashes, and declines in populations of reindeer and caribou. *Ecological Monographs*, 80 (2): 197–219.

Usher PJ (2000) Traditional ecological knowledge in environmental assessment and management. *Arctic*, 53 (2): 183–193.

Usher PJ (2003) Environment, race and nation reconsidered: reflections on aboriginal land claims in Canada. *The Canadian Geographer*, 47 (4): 365–382.

Wake DB and Vredenburg VT (2008) Are we in the midst of the sixth mass extinction? A view from the world of amphibians. *Proceedings of the National Academy of Sciences of the United States of America*, 105 (Suppl. 1): 11466–11473.

White R (1995) *The Organic Machine: the Remaking of the Columbia River.* New York: Hill and Wang.

9 Post-scripting extreme weather

Textuality, eventhood, resilience

Vladimir Janković and James R. Fleming

A small sandy promontory called Willoughby Spit in Norfolk, Virginia, has been a part of the region's landscape for many generations, a residential neighbourhood popular with beachgoers. Its name harks back to Thomas Willoughby, who settled in the area in 1610 and received his first land grant in 1625. His son, Thomas II, as the legend goes, awoke one morning after a powerful storm to see a point of land in front his home, where there had been only water the night before. The Willoughby family applied for an addendum to the original land grant giving them ownership of the 'new' property. More than century later, a hurricane of 1749 washed up an additional eight hundred acres of sand onto the original peninsula, which received its final form during the Great Coastal Hurricane of 1806 (Tyler, 1922).

This is a striking but common story in the history of coastal land formation: a piece of land appears (or disappears) following a hurricane and becomes a fixture in the local landscape (or disappears from it). The spit's existence owes to the sheer violence of nature but its eventual socionatural identity becomes established through human activities, through settlement and property laws and a myriad of social, economic, environmental and aesthetic engagement with the so-called 'natural landscape.' What didn't exist at one time in human history became a 'natural given' at a later time. A hurricane inscribes itself on land as a new spit of land that then becomes a datum of socioenvironmental physiognomy. The spit commemorates a hurricane event while maintaining a mundane reality in the social and geological landscape of the region – it 'stabilizes' an extreme into the everyday (Figure 9.1).

We invoke Willoughby Spit as a parable for how societies come to terms with adverse weather through a *cultural* inscription of extremes, through a discursive and behavioural stabilisation of the extraordinary. Inscriptions of such events by means of the mnemonic vehicles such as local chronicles or commemorations – convincingly explored in the preceding chapters – serve as markers of sociohistorical singularities and contribute new layers of meaning to communities that, by means of such vehicles, assimilate immediate experiences of the extraordinary into a wider horizon of life. Indeed, in the aftermath of disasters, the seemingly miraculous and incomprehensible

Figure 9.1 Willoughby Spit in Norfolk, Virginia.
Source: Pinterest: https://uk.pinterest.com/pin/111041947034832108/ (accessed 2 November 2016). Photo: Vladimir Janković.

acts of nature become stabilised as narratives of suffering, resistance, hope and survival: the socially shared and standardised vehicles designed to re-build community by diverting its gaze from an abyss of destruction and death and towards an understanding of vast inequalities between the banal power of nature and the fragility of human culture.

Episodes of unusual and rare weather, commonly described as severe, extreme, remarkable, exceptional, strange or unique, have been subjects of interest and investigation since the earliest times (Taub, 2003). Such episodes played central roles in world cultures: signifying singular en-counters between societies and forces that challenged them. In mundane parlance, adverse weather disrupts routines and causes damage, discom-fort, delay, illness, loss and death. Yet not all rare and unusual weathers disrupt. Some involve peak experiences of the numinous, the sublime and the memorable: stars twinkling through the powdery falling snow, a personal halo on a cloudbank in the valley below, the rainbow after the storm. This is not to imply that all apprehensions are equal, since the intensity of both disruptions and peak experiences depend on the beholder.

The authors contributing to this book reflect on the sociocultural pro-duction of disruptive weather – on its potential textuality – as each episode that obstructs the expected course of life can become an episode about which something can be said or written. While in principle the weather is always textual ('a sunny day today'), only some weathers count as more than 'phatic' or merely sociable (Golinski, 2007) and thus qualify as sufficiently meaningful to enter the conventional forms of news, obituaries, chronicles, poems and commemorative speeches. Leaving its conventional domain as a mere horizon of life, as an inconsequential signal behind the quotidian, weather, on occasion, foregrounds itself to challenge the routine and shake complacency. From its tacit existence as some 'part of nature', adverse weather turns into social and textual reality, temporarily halting things that have been 'planned', dismantling structures that have been built and inscribing its powers onto the traditions of affected communities. From Pliny to Daniel Defoe to contemporary 'ruin-porn' (Leikam, 2015), writers in both scholarly and lay traditions developed an evocative repertoire to capture this foregrounding. A semiotics of adverse weather has found its voice in classical and 'strange' natural histories, wonder literature, homiletics, magic, science, divination, lore, art and history. The thirteenth-century monk Caesarius of Heisterbach wrote of the judgemental powers of thunderbolts that, in one case, killed a group of 20 men, but spared a priest in their midst: what was the meaning? Bad weather thus occupied the naturalist as much as the cleric, the chronicler and the seer. It transcended cultures, climates and confessions and gave the rationale to providential meteorology, prophetic astronomy, millenarian seismology and apocalyptic hydrology (Heninger, 1978).

Following the themes developed in the preceding chapters, one is led to ponder the circumstances that enable the extraordinary semantic foregrounding of something that, ostensibly, is no more than a violent spike in atmospheric behaviour, an anomalous state of airy affairs? Is weather indeed about the atmosphere, which is a relatively recent conceptualisation of the Aristotelian 'sublunary' world? And if it is not, is it a natural fact or a social artefact, viewed through the lenses of providential history, outdoor nuclear testing or anthropogenic climate change (to name but three)? What makes weather extreme or anomalous – and anomalous in respect to what? How, in the present context, are extremely adverse events magnified (or refashioned) in the acts of retelling and forgetting over time?

Reflecting on the case studies included here, we are struck by the focus on the commemorative ethos of weather knowledge and allusions to a currently unexplored sequence of moments that produce extreme weather cultures. The essays underscore the importance of a retrospective gaze that explains the adherence to norms that constitute the community's weather fate. Such retrospectivity reveals an interest in what had actually happened, in the recognition of a particular weather *event*. In this context, the key moment in the process of coming to terms with the weather impact is one that takes place when a community's unmediated experiences of being 'impacted' by adverse weather begin to emerge as a well-bounded narrative of a singular event. The construction of such an event, from then on, acquires a crucial, organising role in a community's history and memory and its relative place vis-à-vis other communities. This process of event making is not only an instrumental means of a community's coming to terms with concrete impacts and their aftermath. It is also about creating a socially meaningful referent that makes the seemingly unnecessary excesses of nature appear as amenable to textuality and commemorative practices – analogous to the topographers' act of representing Willoughby Spit on a coastal map, an act that solidified the energy of winds and waves into an uncontroversial symbol on the map.

To further illustrate this point, let us briefly turn to the ideas of Jacques Lacan, who sees human existence distributed across the realms of the Imaginary, the Symbolic and the Real. The Imaginary refers to our direct lived experience of reality: how things appear to us. The Symbolic is the realm of structures enabling the Imaginary that allow us to see things the way we see them. But the Real is not a mere external reality; it is rather the 'impossible': something that cannot be meaningfully experienced or symbolised, much like a traumatic exposure to an act of extreme, unfathomable violence that destabilises our entire universe of meaning. The Real, Lacan suggests, can only be identified in its traces, consequences and aftershocks (Žižek, 2014: 120). These traces, effects and aftershocks congeal in the socially shared imaginary of an Event, an element of cognitive inventory that transforms the Real into the Imaginary (in Lacan's terminology). The outcome of this process, the Event, can be reified and thus identified retrospectively – in our context, as a flood, a hurricane landfall or a drought. Such events have life-changing dimensions because they are "the exposure of the reality that nobody wanted to admit (or nobody wanted) but which has now become a revelation and has changed the playing field" (Žižek, 2014: 15).

Fighting high waters in the South of England in 2013 is now known as the Flood of 2013. The flood – that can be seen as made up of a myriad of acts, networks, canals, machines, words, people, institutions, animals, raindrops, freshets and buildings – emerges as the taming of the Real, a discursive sublimation of the unwanted, pointless and unintelligible Real into a singular event with a spatiotemporal meaning. As such, the 2013 flood is given a new lease on life in a range of institutions and practices, where it can now be placed on public and political agendas: the cost of resilience and adaptation, the responsibility for local preparedness, malfunctions in emergency services, linkages between the event and global climate change, the urgency of re-engineering waterways, the need for coastal fortifications, the need to mitigate or intervene in the climate system and the financial concern shared by risk and property industries.

This retrospective production of events is a feature of historicism and as such as can be seen at work in every socially significant situation involving the presence of the Lacanian Real. In this sense, an event – and its subsequent meaning as symbolic of historic 'times' and geographic 'places' – does not need to be judged as destructive only. If anything, the *rarity* of experiences leads to making them exceptional. As the rain returned to the parched British Isles in August 1976 (Waites, this volume) – as the unusually long, hot weather receded into a normalcy of the British damp – the community came to grips with experiences from the 'other side' of exceptionality, from a place and time that allowed one to say "phew" and wonder what, and perhaps why, this had taken place. In the aftermath, it becomes possible to outline contours of the weather 'event' that defied meteorological averages, but also, perhaps more radically, to challenge the British stereotypes, imagine alternative futures or nostalgically re-imagine the summer of 1976 as a one-off event of national significance filtered through the apprehension of particular individuals.

There is a logic to such event forming. As the natural forces unfold during what is to become a weather disaster, one is too close to the scene to discern the full contours of what is taking place. As a soldier in battle knows, the immediacy of the physical assault takes precedence over reflection. As assault shapes a response, the response becomes place and time specific, drawing on the limits imposed by a repertoire of resources. The response depends on material practices and the sedimented rules of coping communities employ to manage distress. With time, however, the experiences of immediate assault gradually take the shape of entities that can be named, initially as minutiae of lived moments of distress but, as time goes on, as formulaic textual inscriptions phrased in a more general and familiar idiom.

To take an imaginary scenario of the processes leading to an emergence of a 'flood': To an individual caught in a heavy rainfall event, a crack in the house appears first as a pang of shock that grows into panic that leads to attempts to shut the hole, to secure belongings, to make decisions about whether and when to leave, and if so, what to wear, what valuables to take, who to call for help, and where to go. While this maelstrom of thoughts, feelings, fears, dilemmas, decisions and pains rush through one's mind in circumstances that preclude their naming, the resultant product, at a later time, becomes legible as a series of multiple, serialised and superimposed events-in-the-making: the 'moment we discover the leak' and the 'time when we left home', which, in subsequent official reports, are phrased in the evermore-general terms of handling of public safety, evacuation, displacement and damages. Eventually, from a more distant perspective, the subjectivities of suffering and individual decisions solidify into a single master event of the 'flood.'

The flood thus appears not when and where it hits, but only when the acts of coping forge into a narrative of survival that, from another, safer and more distant perspective, acquires a meaning as a unitary event – a disaster. Much like accounts of major battles, visceral experiences that could not initially be grasped or dealt with or systematised at the time of distress, have now been replaced by an entity with a known date, location, duration and effects, an entity situated in the framework it shares with non-extreme, routine events of mundane experiences. The Real has been solidified through 'translational' acts of coping that inscribe the seemingly wanton excess of nature into the fabric of meaning, the process in which environmental emergency becomes normalised and further elaborated through acts of commemoration.

The subjectivities of hardships, textually sublimated as events, are maintained and cultivated as identity markers of the sociopolitical makeup of communities through mnemonic practices. Both tangible and intangible forms of cultural heritage are enjoined in these practices: from statues, memorial sites to folkloric traditions, poetry and prose, acts of re-enactment and remembering configure 'original' events as self-standing markers that contribute to intergenerational historic identity (Hall and Endfield, 2015) and collective memory (Barnier and Sutton, 2008). But how do these acts of commemoration affect the society in which they occur? Is there a 'purpose' to these acts, as it were, a purpose in the sense of an evolutionary mechanism that a collective mind develops to assuage pain by memory? Or a purpose in the sense of a Douglasian normalisation of an event 'out of place' that is returned to its place by the textual configurations of jarring experiences? What logic explains the cultural practices of remembering something that is best forgotten? What benefit is there in dwelling on past traumas? Or might they reoccur?

There is a sense of inevitability here: just as a hurricane can 'invent' a piece of land that becomes a new reality, so a cultural experience of hurricanes, floods, storms and other out-of-the-ordinary events leaves in its wake a spate of cultural inventions in the representational spaces of poetry, history, memory, drama and visual arts. There is a parallel between weather's ferocity and its ferocity over human souls. When a tornado passes through a heavily wooded area, many trees that were not felled outright have their tops heavily pruned, giving them a 'bonsai' appearance. These features can linger for many decades, for the remaining life of those trees. Sometimes a 'swath' of new growth is all that remains as evidence of the tornado's passage, as if a giant's logging road was now growing over.

As a tornado inscribes its path in shattered trees, its victims inscribe the tornado in words and things: these bruises heal and become painless in time, but they also remain visible to the eyes perusing the chronicles, the mirrors of history. That a society will 'commemorate' an event that it would rather forget is in the nature of inevitability: as Willoughby Spit lives on as a 'thing of the world', so do mementos of distress persist in memory and shape identity beyond original trigger events. These mementos are 'things of the world', outlining wider contours of environmentality. With them, the horizon of the possible is expanding as is social strength in coping with future distress. It makes sense to say that, as Morgan (2015) suggests, early modern societies encoded their relationships to water in a number of genres as a way of coping with flood impacts. These forms and encodings also represent attempts to come to terms with ever-more-challenging futures, expanding the inventory of environmental cultures. Whether physical, oral, scribal or artistic, markers of weather shocks represent these shocks as 'history' to simultaneously vindicate life that, with new misfortunes, is further strengthened through endurance, solidarity, mercy and sacrifice. These markers announce good fortunes for those who reminisce and who thus assert their defiance to chance acts of the Real.

Ostensibly, commemorating extreme weather is about reaffirming power over grief; but it is, more significantly, about those who commemorate and reaffirm this power, those who in these acts secure their dignity vis-à-vis that which they remember. It is about those who through thankfulness can re-turn, re-think, and re-member events that once dis-membered their fabric of life. These acts, as some of the authors in this collection indicate, suggest that the presence of raw memories can prevent early anniversaries of disasters, highlighting the importance of the *distance* in accomplishing the closure of agonising experiences through the palliative textuality of mnemonic inscriptions – i.e. accomplishing a transposition of the unbearable workings of Lacan's Real into the convention of the oral and written word, documentary archives and physical monuments.

These therapeutic re-enactments of distress imply the psychological importance of the temporal distance from the events, the distance that allows for a reflective, quasi-dispassionate remembering of what once was a fresh, burning wound on the tissue of social life. Re-enactments help the wound turn into a scar. Consider the seventeenth-century Dutch 'shipwreck' paintings that attracted (and still attract) viewers' attention partly because of their striking iconography and their framing of distress, but partly because these paintings remind the spectator of a distance from which he or she is now allowed to see voyeuristically the wreckage as somebody else's tragedy, at the same time entertaining a pious moral choreography of sympathy, sorrow and respect for the now unknown and unknowable victims. The people on board are always 'others' in distress, long dead or unknown for whom, from a distance of spectatorship, one has the privilege to indulge in a moral economy of compassion. In this way, painterly commemoration of shipwrecks casts the spectator's own condition in sharp, reassuring relief against the condition of those in agony, giving rise to a dispassionate reflection on danger and destruction in the safe haven of the aesthetic experience, not to mention the profoundly didactic potential that such events, and shipwreck events in particular, may have in the spiritual life of a religious community (Janković, 2009).

By means of distance, difference and contrast, the re-enacting of a punctuated equilibrium reaffirms the fundamental importance of the 'baseline' balance between nature and society – if we can be temporarily excused for making such a problematic dichotomy. From high antiquity, philosophers and theologians regarded providential order in both its ordinary and extraordinary manifestations. In what is known as the Virgilian (Jupiter) Theodicy, natural disasters are God's tests to human ingenuity. With each disaster, God gives humans a chance to respond, enriching their inventory of coping practices that, in due course, result in the growth of civilization itself (Miles, 1980).

There is then, we suggest, a compensatory logic to *preserving* and *curating* weather disasters: unless acknowledged and documented, owned and assimilated, remembered and re-enacted, disasters signify nothing, remaining purposeless blips in the course of brute nature, inflicting unnecessary pain on the unsuspecting human tribe. Suffering, coping and restitution of normalcy would, in a pre-textual world, signify nothing other than a wanton effort spent on rectifying futility of natural excess. This hypothetical 'nothingness' of disasters, however, their seeming absurdity, their seeming challenges to logos and Divine order, their utter *uselessness*, would mean an infinitely more painful tragedy than physical suffering itself. But assuming uselessness and burying memories is demoralising and impious: for disaster must have a meaning and purpose to make surviving them worthwhile. Their mementos inscribe a dignity and admiration, a social cohesion that upholds the continuity of life into the future.

We opened our discussion with an allegorical imagery of Atlantic hurricanes as geological 'scribes' of coastal landscapes. We wish to conclude with two examples that complement this volume's rich explorations into curatorial traditions and weather memories. The first one is the great flood of 1889 that effectively wiped the city of Johnstown, Pennsylvania, off the map. The flood was caused by the catastrophic failure, after heavy rains, of the South Fork dam on the Little Conemaugh River, some 12 miles upstream. At the time, Johnstown was a working-class, steel company town with a population of 30,000. The flood killed over 2,209 people and completely devastated the city's infrastructure (McCullough, 1968). Other significant floods occurred in 1894, 1907 and 1924, but the next great deluge occurred around Saint Patrick's Day 1936 during a two-week period of almost continuous rain that raised the river level 17 feet and killed 25 people. Some 40 years later, on 19 July 1977, a very strong overnight storm pounded the watershed area above Johnstown, and the nearby rivers began to swell in an eerie reminder of the flood of 1889. Six dams in the area failed, and in the morning, Johnstown was under 8 feet of water and 80 people had lost their lives. As of 2016, the 'Flood City' is still devastated, with no industrial base and a population some 10,000 less than it was 127 years earlier.

Antediluvian Johnstown is *separated* from the Johnstown of 1889 by the profound epistemic rupture of the great flood. Since then, its citizens have been united by shared common experiences of loss and recovery. There are currently two Johnstown Flood-related commemorative sites in the area. Johnstown Flood National Memorial preserves the remnants of the South Fork Dam. The Memorial staff runs daily screenings of a National Park Service-produced film, *Black Friday* that retells the 1889 tragedy. In the city of Johnstown, an old Carnegie Library houses the Johnstown Flood Museum, operated by the Johnstown Area Heritage Association, which features the *Johnstown Flood,* an Academy-Award-winning film by Charles Guggenheim (Figure 9.2).

Our second story comes from Austrian Tyrol. At 3:59 pm on February 23, 1999, an extraordinarily destructive avalanche descended on the popular Alpine skiing resort of Galtür, arriving in the village from the mountain region officially designated as a safe (green) avalanche zone. Triggered by almost four meters of fresh snow and strong winds, the 50-meter-high wall of

Figure 9.2 Detail of The Great Conemaugh Valley Disaster – Flood & Fire at Johnstown, PA., subtitled Hundreds Roasted Alive at the Railroad Bridge.

Source: Reproduced from a lithograph print published by Kurz & Allison Art Publishers, Chicago, Illinois. 1890; Wikimedia Commons, https://commons.wikimedia.org/wiki/File:The_Great_Conemaugh_Valley_Disaster.jpg (accessed 2 November 2016).

powder took one minute to reach the resort with the speed of 290 kilometers per hour, destroying dozens of buildings and depositing 170 thousand tonnes of snow. It killed 31 people. The Austrian government called for international help, and thousands of tourists and local residents were airlifted out of the area. Local residents demanded to know why the avalanche arrived from a safe zone, but hazard zoning could be made only on the basis of the historical record, which in this case contained no information of avalanches coming from the said area. The event was widely reported by the global media and has recently been examined within the anthropological context of the Alpine 'disasters tradition' (Hinrichsen, 2015) (Figure 9.3).

Our interest here is in Alpinarium Galtür, the village multimedia museum dedicated to Alpine communities and a memento to the 1999 catastrophic avalanche. The Alpinarium consists of an elongated building connected to a stone wall, both of which lie perpendicular to the direction of the 1999 avalanche. The architectural merging of the museum and the wall suggest a protective shell, an architectural rendition of Galtür's disaster and the Alpine disaster tradition more generally. The conscious merger of the architectural form, local building materials embedded in the wall and the art/historic content of exhibition space shown in a modernist interior capture the hybridity of the discursive elements we discussed above, especially the retrospective restoration of the trigger event as a marker of community's

Figure 9.3 Alpinarium Galtür: the 345-m-long and up to 19-m-high avalanche barrier incorporates art and culture; interior design and an exterior integrate
seamlessly with the natural surroundings of the Paznaun valley in Tyrol.
Photo: Vladimir Janković.

history and identity, symbolically and physically acting as a reminder of,
and a barrier against, future disasters. The Alpinarium stabilises natural
violence by combining the physical and the curatorial expression of grief
and resilience, thus integrating the extraordinary into the everyday landscape of Paznaun Valley (Figure 9.4).

As the chapters presented in this volume bring in sharp focus, the commemorative construction of weather events often sees material culture
'partnering' with social reality and textual imaginaries to produce lasting
monuments of meteorological stress experienced by world communities. We
conclude by noting that such creative reconstructions also create an organic,
powerful sense of solidarity. Caught in a severe weather episode and cornered
into a claustrophobic sense of imminent danger, people experience a severe
and humiliating reduction of freedom and dignity. But they compensate this
physical, external helplessness through a conscious, ecstatic effort to expand
their *inner* space, creating a substitute for the physical pain and loss that
they cannot overcome by means commensurable to that of cosmic elements.
Endurance, voluntarism, sacrifice and solidarity triumphantly emerge from
these diminished spaces, to prevail over their force and reclaim the temporarily confiscated freedom that, with time, is memorialised as identity, resilience and ingenuity crafted during the community's most trying times.

Figure 9.4 The permanent Alpinarium exhibit features a collection of spinning top-shaped wooded artefacts whose plates' diameters represent annual snowfall in different locations in Tyrol.

References

Barnier AJ and Sutton J (2008) From individual to collective memory: theoretical and empirical perspectives. *Memory*, 16 (3): 177–182.

Golinski J (2007) *British Weather and the Climate of Enlightenment*. Chicago: Chicago University Press.

Hall A and Endfield GH (2015) "Snow scenes": exploring the role of memory and place in commemorating extreme winters. *Weather, Climate, and Society*, 8 (1): 5–19.

Heninger SK (1978) *A Handbook of Renaissance Meteorology*. Durham, NC: Duke University Press.

Hinrichsen J (2015) Avalanche catastrophe and disaster traditions: anthropological perspectives on coping strategies in Galtür, Tyrol. In Gerstenberger K and Nusser T (eds.) *Catastrophe and Catharsis: Perspectives on Disaster and Redemption in German Culture and Beyond*. Camden House: Boydell and Brewer: 155–171.

Janković V (2009) The Wind Euroclydon. Climatological hermeneutics and environmental individuation from Bishop Bentley to Captain Smith. In Nova A and Michalsky T (eds.) *Wind und Wetter: Der Ikonologie der Atmosphäre*. Rome: Marsilio: Marsilio.

Leikam S (2015) Picturing high water: the 2013 floods in Southeastern Germany and Colorado. In Leyda J and Negra D (eds.) *Extreme Weather and Global Media*. London: Routledge: 74–99.

McCullough D (1968) *The Johnstown Flood*. New York: Simon and Schuters.

Miles GB (1980) *Virgil's Georgics: A New Intepretation*. Berkeley, CA: University of California Press.

Morgan RA (2015) *Running Out? Water in Western Australia* Crawley. Western Australia: UWA Publishing.

Taub L (2003) *Ancient Meteorology*. London: Routledge.

Tyler LG (1922) *History of Hampton and Elizabeth City County Virginia*. Hampton, VA: The Board of Supervisors of Elisabeth City.

Žižek S (2014) *Event*. London: Penguin.

Index